Advancing Global Bioethics

Volume 14

The book series Global Bioethics provides a forum for normative analysis of a vast range of important new issues in bioethics from a truly global perspective and with a cross-cultural approach. The issues covered by the series include among other things sponsorship of research and education, scientific misconduct and research integrity, exploitation of research participants in resource-poor settings, brain drain and migration of healthcare workers, organ trafficking and transplant tourism, indigenous medicine, biodiversity, commodification of human tissue, benefit sharing, bio-industry and food, malnutrition and hunger, human rights, and climate change.

More information about this series at http://www.springer.com/series/10420

Joris Gielen
Editor

Dealing with Bioethical Issues in a Globalized World

Normativity in Bioethics

Springer

Editor
Joris Gielen
Center for Healthcare Ethics
Duquesne University
Pittsburgh, PA, USA

ISSN 2212-652X ISSN 2212-6538 (electronic)
Advancing Global Bioethics
ISBN 978-3-030-30431-7 ISBN 978-3-030-30432-4 (eBook)
https://doi.org/10.1007/978-3-030-30432-4

This Springer imprint is published by the registered company Springer Nature Switzerland AG.
The registered company address is: Gewerbestrasse 11, 6330 Cham, Switzerland

Contents

About the Authors

Kiarash Aramesh is assistant professor in the Department of Biology and Health Sciences at Edinboro University of Pennsylvania (EUP). He also serves as the director of the EUP's James F. Drane Bioethics Institute. He is a physician with specialty in community medicine. He also holds a Ph.D. in Healthcare Ethics. Before joining the EUP, he worked as a visiting scholar at the Center for Healthcare Ethics at Duquesne University (2017–2018) and at the Department of Bioethics at the National Institutes of Health (NIH) (2013–2014). He also worked as a faculty member at the Medical Ethics and History of Medicine Research Center at Tehran University of Medical Sciences from 2005 to 2018. He held this position at the time of writing this chapter.

Pascal Borry is associate professor of Bioethics at the Centre for Biomedical Ethics and Law, University of Leuven, Belgium. His main research activities are concentrated on the ethical, legal, and social implications of innovative technologies. He published more than 150 publications in international peer-reviewed journals on topics such as direct to consumer genetic testing, public health genomics, biobanking, preconceptional and prenatal screening, neonatal screening, and anti-doping. He is program director of the Master of Bioethics. Within the European Society of Human Genetics, he was member of the Professional and Public Policy Committee (2009–2016) and elected member of the board (2012–2017). Since 2016, he is member of the Ethics Panel of the World Anti-Doping Agency.

Jonathan K. Crane serves as the Raymond F. Schinazi scholar in Bioethics and Jewish Thought in the Center for Ethics at Emory University, where he is also an associate professor of Medicine and of Religion. His previous books include *The Oxford Handbook of Jewish Ethics and Morality* (coedited with Elliot Dorff), *Narratives and Jewish Bioethics*, *Beastly Morality: Animals as Ethical Agents*, and *Eating Ethically: Religion and Science for a Better Diet*. He is a coeditor in chief of the *Journal of Jewish Ethics*.

Joris Gielen is director of the Center for Healthcare Ethics at Duquesne University in Pittsburgh. He studied history, religious studies, and theology at the University of Leuven (Catholic University of Leuven, Belgium) and Indian philosophy at Banaras Hindu University, Varanasi, India. In his research, he focuses on religion, spirituality, and ethics in palliative care from a global perspective. He has conducted research projects in Belgium, the USA, and India. His special area of interest is the role of religion in the experiences of patients, their family members, physicians, and nurses in palliative care in North India and how religion and spirituality influence medical decision-making. Currently, he is collaborating to a project sponsored by the Indian Council of Medical Research that aims at developing a scale to measure spiritual distress in Hindi-speaking palliative care patients in India.

Soraj Hongladarom is professor of Philosophy and director of the Center for Ethics in Science and Technology at Chulalongkorn University in Bangkok, Thailand. He has published books and articles on such diverse issues as bioethics, computer ethics, and the roles that science and technology play in the culture of developing countries. His concern is mainly on how science and technology can be integrated into the lifeworld of the people in the so-called Third World countries and what kind of ethical considerations can be obtained from such relation. A large part of this question concerns how information technology is integrated in the lifeworld of the Thai people and especially how such integration is expressed in the use of information technology in education. He is the author of *The Online Self: Externalism, Friendship and Game* and *A Buddhist Theory of Privacy*, both published by Springer. His articles have appeared in *The Information Society, AI & Society, Philosophy in the Contemporary World,* and *Social Epistemology,* among others.

Fahmida Hossain is a Ph.D. student in the Center for Healthcare Ethics and a graduate assistant at the Simon Silverman Phenomenology Center at Duquesne University. Her vision is to develop narrative and storytelling as a lived and practical means in healthcare to understand and shape organizational culture, change efforts, and improve best practices everyday routine, and person-to-person encounters through an innovative approach she calls narrative authority. She also serves as a training facilitator and content developer for the narrative-focused, coaching and organizational change firm, Naridus, LLC. She has co-facilitated development classes for the Mandela Washington Fellowship, part of President Obama's Young African Leaders Initiative; the Johnson Institute for Public Leadership; and the Health Career Scholars Academy. She held two research fellowships at the Jewish Healthcare Foundation in the areas patient safety fellowship and death & dying. She holds a master's degree in Leadership, Professional Administration from Duquesne University. She is originally from Dhaka, Bangladesh. She is based in Pittsburgh, PA with her husband and their young son.

Kenneth Kipnis is professor emeritus of Philosophy at the University of Hawaii at Mānoa, having taught medical ethics there for 35 years and retiring in 2016. He has published broadly on ethics in pediatrics, legal practice, prison and military medicine, criminal justice, research ethics, and disaster medicine. He has served on several hospital ethics committees, drafted health-related legislation, developed ethics curricula for the University of Hawaii's schools of medicine and nursing, and worked regularly as an expert witness in ethics-related court cases. He chaired the Committee on Philosophy and Medicine of the American Philosophical Association, served as a Visiting Senior Scholar at the American Medical Association's Chicago headquarters, and has been a visiting senior faculty member at the University of Texas Medical Branch in Galveston as well as at the College of Charleston in South Carolina.

Gerard Magill graduated in 1987 with his Ph.D. degree from Edinburgh University in Scotland. From 1987, he developed his scholarly career at Saint Louis University where in 1996 he was appointed as the Department Chair of the Center for Healthcare Ethics in the University's Health Sciences Campus. In 2007, he moved to Duquesne University in Pittsburgh where he holds the Vernon F. Gallagher Chair as a tenured Professor in the Center for Healthcare Ethics. He has authored, co-authored, or edited 11 books including a co-authored textbook on health care ethics and a forthcoming co-authored book on governance ethics for boards of directors in health care. He has published many scholarly and professional articles, is a member of 15 professional associations, and specializes in institutional review boards, hospital ethics committees and ethics consultation services, and human genomics.

Rouven Porz is head of the Clinical Ethics Unit of the University Hospital in Bern, Switzerland. He is also the president of the EACME (European Association of Centres of Medical Ethics). He studied biology, philosophy, and pedagogy. He conducted both empirical and conceptual research in the field of genetic testing and reproductive medicine before he stared focusing on the implementation of clinical ethics in health care setting. He lives in Bern, is married to a neurosurgeon, and spends his free time with his daughter and his dog "kairos."

Henk ten Have is professor emeritus of Healthcare Ethics in the Center for Healthcare Ethics at Duquesne University in Pittsburgh, USA. He studied medicine and philosophy in the Netherlands and worked as professor in the Faculty of Medicine of the Universities of Maastricht and Nijmegen. From 2003 until 2010, he has joined the UNESCO in Paris as director of the Division of Ethics of Science and Technology. His latest books are *Global Bioethics: An Introduction* (2016), *Vulnerability: Challenging Bioethics* (2016), and *Wounded Planet: How Declining Biodiversity Endangers Health and How Bioethics Can Help* (2019).

Chapter 1
Introduction: Normativity in Context

Joris Gielen

Keywords Normativity · Religion · Normative ethics · Pluralism · Globalization · Healthcare · Global bioethics · Empirical research

In bioethical publications and debates, the concept of normativity is most often used without consideration of the difficulties surrounding it. This is in fact surprising as there are many competing claims for normativity within bioethics. We also cannot merely assume that a Western understanding of normative bioethics will be unproblematic in bioethics in non-Western cultures and religions. These fundamental problems challenge the current understanding of normativity in bioethics and need to be addressed in order to be able to understand and develop appropriate answers to bioethical issues in a globalized world with multicultural societies.

In the 2009 edition of their influential textbook *Principles of Biomedical Ethics,* Tom L. Beauchamp and James F. Childress distinguish between "normative ethics" and "nonnormative ethics". Normative ethics is of two types: "general normative ethics" and "practical ethics". Ethicists focusing on general normative ethics attempt to determine "[w]hich general moral norms for the guidance and evaluation of conduct should we accept, and why." "Practical ethics" refers to every "attempt to interpret general norms for the purpose of addressing particular problems and contexts" (Childress and Beauchamp 2009, pp. 1–2).

This straightforward description of normative ethics, which Beauchamp and Childress subsequently apply to bioethics, initially leaves two important questions unanswered. The first question is how people can deal or should deal with a plurality of bioethical discourses that all claim normativity. Examples of such competing normative bioethical discourses can be perceived in variations and differences in bioethical arguments within individual religions, and the opposition between bioethical arguments from specific religions and arguments from bioethicists who do

J. Gielen (✉)
Center for Healthcare Ethics, Duquesne University, Pittsburgh, PA, USA
e-mail: gielenj@duq.edu

J. Gielen (ed.), *Dealing with Bioethical Issues in a Globalized World*, Advancing Global Bioethics 14, https://doi.org/10.1007/978-3-030-30432-4_1

not claim religious allegiance. The second question is to what extent and in what form normative bioethics can also be found in other cultures and societies, because Beauchamp and Childress primarily described normative ethics within the context of Western bioethics. These are pressing questions, as today bioethical issues present themselves all over the world, in a wide variety of cultural settings. Almost everywhere, cultural and religious diversity has become an important part of medical reality. This diversity presents a challenge to the development of bioethical arguments, also in the West.

In a way, normativity has always been a problematic concept within ethics. The first chapters of this book, on normativity in religious bioethics, will illustrate this. These chapters will show that within individual religious traditions the normativity of bioethical arguments has always been contested. Within these traditions, there has been a variety of bioethical views that all claim to be normative. As society has changed, the problems surrounding normativity in ethics have not only become more complicated, but also more acute. We are now living in a global society and are, also, talking about global bioethics. From the perspective of global bioethics, normativity is an important issue.

People are moving throughout the world like never before. In search of job opportunities and a better life, people migrate from developing countries to developed countries and when they travel, they take their worldviews and ethical attitudes with them to their new homes. When these people need healthcare, they are confronted with a healthcare workforce that does not necessarily share their ethical views. Driven by a shortage of healthcare professionals in developed countries, healthcare professionals from developed countries move to the West where they are confronted with ethical attitudes they may not share. When an ethical issue arises, they may get into conflict with their colleagues in healthcare. At the same time, healthcare itself has become a global phenomenon as exemplified in issues such as international medical research, and medical tourism. In all these situations, when conflict arises, the question is which view should prevail? How to reach consensus? Which ethical argument has normative authority? Within the context of a globalized world and global bioethics, the problems of normativity have, indeed, become more acute.

In this book, we will analyze the ethical diversity that exists in healthcare in a globalized world and try to understand how it is still possible to develop normative ethical arguments in such a context. The emphasis of this book is practical rather than philosophical. We will, for instance, look at the religious ethical diversity that healthcare providers may see in practice. But, the solutions, too, will be practical and aimed at presenting ways to overcome or prevent problems related to competing normative ethical discourses in healthcare.

Chapters 2–7 constitute the first part of the book, which, as already indicated, deals with normative bioethics in religious traditions. How do problems regarding normative bioethics surface within Hinduism, Buddhism, Islam, Judaism, and Roman Catholicism, and are there ways to overcome these problems formulated within these traditions? These chapters should, in no way, be seen as an attempt to cover the entire spectrum of religious bioethics. There are many religious traditions or forms of religiosity and spirituality that these chapters do not cover and the religious answers to concrete bioethical issues that are discussed are only examples intended to illustrate that particular religion's approach to normative bioethics. The focus of the chapters is, indeed, on the method of normative bioethical enquiry.

In that first part, after a very brief introduction to that part, we turn our attention towards Islam. When Western healthcare scholars study Islam (Chap. 4) they too easily forget the diversity that exists in this religion (see e.g. Gilliat-Ray 2003). This diversity also perspires in bioethical views. In this regard, the division between Shī'ite and Sunnī Muslims is only the most visible sign of intra-Islamic bioethical diversity (Brockopp and Eich 2008; Sachedina 2005). Similar issues arise upon critical examination of normative bioethical discourse within Judaism (Chap. 4) and Roman Catholicism (Chap. 5). Jewish ethicists have reached diverging conclusions regarding important bioethical issues such as euthanasia (Baeke et al. 2011). The Roman Catholic magisterium, and theologians and scholars have expressed their views on bioethical issues. Sometimes, these views are at odds (Peppin et al. 2004, pp. 1–130).

In Asian Hinduism (Chap. 6) and Asian Buddhism (Chap. 7) the comparatively limited amount of extensive bioethical publications may even make scholars wonder whether normative bioethics has an important place in these religions or exists at all. Most substantial works that claim to represent normative bioethics from a Hindu or Buddhist perspective have been written by academic scholars who live in the West and are not necessarily adherents to these religious traditions (e.g. Crawford 1995, 2003; Keown 2001). However, this does not exclude the possibility that Asian Hindus and Buddhists do feel compellingly guided by their tradition when they take biomedical decisions. In this way, in Asian Hinduism and Buddhism, normative bioethics may even exist in another form than systematic publications (Clooney 1995).

In the second part (Chaps. 8–13), normativity at crossroads, the discussion is broadened to normative bioethics outside an explicitly religiously committed context, and specific attention is paid to bioethics as an interdisciplinary endeavor particularly within the context of healthcare practice. Also from this perspective, normative bioethics is facing challenges that may force scholars to reconsider the meaning and scope of normativity. These challenges can be perceived in bioethical research, global bioethics and the challenges posed by the particularity of contemporary Western society, which is at same time secular and pluralistic.

In bioethical research (Chap. 9) the 'empirical turn in bioethics' has opened up new avenues of research in the past two decades but has also forced ethicists to ponder upon the complex relationship between empirical and normative bioethics (Borry et al. 2005; Goldenberg 2005; Hurst 2010; Kon 2009; Leget et al. 2009).

Global bioethics (Chap. 10) has arisen in the context of the spread of Western medicine due to which bioethical challenges manifest themselves all over the world. Moreover, because of processes of globalization, the consequences of bioethical issues very concretely link developed and developing countries. In this regard, reference can be made to phenomena such as medical tourism, international surrogacy and drug trials conducted in developing countries but mainly for the long-term benefit of patients in developed countries. In response to these evolutions global bioethics has developed. Yet, ethicists have wondered to what extent normative global bioethics, which transcends cultures, is a sustainable endeavor (ten Have and Gordijn 2014).

All issues mentioned above converge in contemporary Western society, which is secular and pluralistic but at the same time also deeply influenced by its own historical, religious and philosophical roots (Chap. 11). In such society, healthcare providers are confronted with various, sometimes conflicting, normative bioethical discourses. E.g., a physician using a very secularized ethical language might be confronted with a person strictly adhering to religious norms directly linked to a specific religious tradition. Driven by these norms the patient may reach conclusions regarding refusal or demand of treatment that are very different from the physician's. This generates the question how health care professionals should respond to this variety (Bracanovic 2011) and whether meaningful dialogue between competing normative perspectives is possible in healthcare. It, also, triggers the question what healthcare organizations should do, in this regard to support their employees and patients (Chap. 12). An important place in that dialogue may be reserved for ethics consultants, provided they have the necessary authority to lead (Chap. 13).

Each individual chapter is intended as an introduction to its topic aimed at a broader audience. Readers who are interested to gain more profound insight into these topics will find suggestions for further reading at the end of each chapter.

References

Baeke, G., J.P. Wils, and B. Broeckaert. 2011. 'There Is a Time to Be Born and a Time to Die' (Ecclesiastes 3:2a): Jewish Perspectives on Euthanasia. *Journal of Religion and Health* 50 (4): 778–795.

Borry, P., P. Schotsmans, and K. Dierickx. 2005. The Birth of the Empirical Turn in Bioethics. *Bioethics* 19 (1): 49–71.

Bracanovic, T. 2011. Respect for Cultural Diversity in Bioethics. Empirical, Conceptual and Normative Constraints. *Medicine, Health Care, and Philosophy* 14 (3): 229–236.

Brockopp, J.E., and T. Eich, eds. 2008. *Muslim Medical Ethics: From Theory to Practice*. Columbia: University of South Carolina Press.

Childress, J., and T. Beauchamp. 2009. *Principles of Biomedical Ethics*. New York/Oxford: Oxford University Press.

Clooney, F.X. 1995. Back to the Basics: Reflections on Moral Discourse in a Contemporary Hindu Community. *The Journal of Medicine and Philosophy* 20 (4): 439–457.

Crawford, S.C. 1995. *Dilemmas of Life and Death: Hindu Ethics in a North American Context*. New York: State University of New York Press.

———. 2003. *Hindu Bioethics for the Twenty-First Century*. Albany: State University of New York Press.

Gilliat-Ray, S. 2003. Nursing, Professionalism, and Spirituality. *Journal of Contemporary Religion* 18 (3): 335–349.

Goldenberg, M.J. 2005. Evidence-Based Ethics? On Evidence-Based Practice and the "Empirical Turn" from Normative Bioethics. *BMC Medical Ethics* 6: E11.

Hurst, S. 2010. What 'Empirical Turn in Bioethics'? *Bioethics* 24 (8): 439–444.

Keown, D. 2001. *Buddhism and Bioethics*. New York: Palgrave.

Kon, A.A. 2009. The Role of Empirical Research in Bioethics. *The American Journal of Bioethics* 9 (6–7): 59–65.
Leget, C., P. Borry, and R. de Vries. 2009. 'Nobody Tosses a Dwarf!' The Relation Between the Empirical and the Normative Reexamined. *Bioethics* 23 (4): 226–235.
Peppin, J.F., M.J. Cherry, and A. Iltis, eds. 2004. *Religious Perspectives in Bioethics*. London: Taylor and Francis London.
Sachedina, A. 2005. Islamic Ethics: Differentiations. In *The Blackwell Companion to Religious Ethics*, ed. W. Schweiker, 254–267. Malden/Oxford/Victoria: Blackwell Publishing.
ten Have, H.A.M., and B. Gordijn. 2014. Global Bioethics. In *Handbook of Global Bioethics*, ed. H.A.M. ten Have and B. Gordijn, vol. 1, 3–18. Dordrecht: Springer.

Part I
Normative Bioethics in Religious Traditions

Chapter 2
Introduction to Part I: Normative Bioethics in Religious Traditions

Joris Gielen

Keywords Islam · Judaism · Roman Catholicism · Hinduism · Buddhism · Bioethics · Normativity · Religious diversity

The five chapters on religious bioethics show how Islam, Judaism, Roman Catholicism, Hinduism and Buddhism have developed normative answers to bioethical issues. As explained before, throughout these chapters, the emphasis is not on the religious traditions' concrete responses to particular bioethical issues. Whenever such responses are analyzed in the chapters, this is part of the analysis of that religious tradition's approach to normative bioethics in general. The descriptions make clear that within each of the studied religious traditions, the adopted method towards normative bioethics, or, sometimes, a plurality of methods, has nurtured bioethical diversity within these traditions. Still, that diversity has not led the traditions to the conclusion that normative bioethics is impossible. In most instances, within religious traditions, diversity rather seems to have acted as a stimulation to further refine both concrete responses to particular bioethical issues as well as fundamental development of the basis or methods that sustain these arguments.

In the chapter on normativity in Islamic Bioethics, Kiarash Aramesh shows that the debate on normativity is nearly as old as Islam itself. Discussion on the position of reason within normative ethical discourse led to the development of different methods to resolve ethical questions. This discussion became intertwined with the question to what extent Islamic jurisprudence can or should provide an answer to bioethical issues.

In the chapter on normativity in Jewish bioethics, Jonathan Crane argues that normativity is accepted as a category within Judaism, although its content is

J. Gielen (✉)
Center for Healthcare Ethics, Duquesne University, Pittsburgh, PA, USA
e-mail: gielenj@duq.edu

J. Gielen (ed.), *Dealing with Bioethical Issues in a Globalized World*, Advancing Global Bioethics 14, https://doi.org/10.1007/978-3-030-30432-4_2

contested. Using the example of Jewish discourse on death and dying, Jonathan Crane shows the disagreements regarding definitions, interpretation of classic texts, and the application of foundational worldviews. Within Judaism, there is, also, disagreement regarding the authoritative position of rabbis. Jonathan Crane concludes that, notwithstanding all these disagreements, Jews continue trying to construct norms that are applicable to the modern biomedical context, but, given the diversity of approaches to the tradition, the outcome of that endeavor is manifold.

Contrary to Judaism, Roman Catholicism, which is just one branch of the Christian tradition, has a strong organizational structure that has enabled it to develop streamlined and coherent doctrinal teaching. This does not mean, however, that there is no discussion or dissent within the Roman Catholic tradition. In the chapter on normativity in Roman Catholic bioethics, Gerard Magill describes how the imagination can nurture bioethical rational discussion. The imagination can challenge existing conclusions and direct reasons to converge towards a new conclusion. In his argument, Gerard Magill draws inspiration from the nineteenth-century Roman Catholic theologian John Henry Newman and he illustrates his point with the principle of double effect, which has a long tradition in Catholic moral theology.

The Asian religions of Hinduism and Buddhism show very different perspectives on how to make sense of normative bioethical diversity. In Hinduism, diversity is so extensive that some scholars have even wondered whether there really exists something like a Hindu religion. The chapter on Hindu bioethics argues that it, nevertheless, makes sense to speak about normativity in Hindu bioethics. Normative bioethical discourse in Hinduism is not primarily to be found in systematic publications, but, rather, in the views and opinions of Hindus who are confronted with bioethical issues in their daily lives, such as physicians and nurses. Normative arguments in Hindu bioethics are developed through an approach that combines virtue ethics and deontological ethics. The chapter shows that by doing so, Hindu bioethics is able to integrate various perspectives and avoids the pitfalls of both ethical absolutism and relativism.

In the chapter on normativity in Buddhism, Soraj Hongladarom shows that Buddhism succeeds in avoiding that pitfall through – what he calls – "ethical teleological naturalism." The author explains that, in Buddhist bioethics, the emphasis is not on the content of norms, but on the ultimate Goal towards which these norms should lead: *Nirvāṇa.* Consequently, depending on the context, Buddhist responses to an ethical issue such as euthanasia could vary as long as the response brings the individual closer to the ideal of *Nirvāṇa.* This approach, indeed serves as an antidote to ethical absolutism through the emphasis of the contextuality of norms. Simultaneously, Buddhist bioethics rejects relativism by upholding the universality of the Goal.

These five chapters on religious bioethics do show two things. First, they illustrate the intrareligious diversity in bioethical views within each religion. Ethicists within the studied religions have struggled to agree on normative bioethical positions. So, if there cannot be agreement within worldviews, it would be tempting to conclude that agreement will remain forever elusive and that moral relativism is the only viable position. This is precisely where the second point comes in. The chapters also show that the validity of a claim to normativity within a religion does not depend on whether or not that claim is dominant within that religion or the only such claim. Religious normativity rather seems to reside in the process that is followed in argument development. The chapters will show the different processes that are followed within the religions: some degree of respectful consideration of tradition, use of the imagination, a combination of ethical theories, or ethical teleological naturalism. What all these approaches do have in common is that they allow ample scope for plurality and dissent.

Chapter 3
Normativity in Islamic Bioethics

Kiarash Aramesh

Abstract In Islam, debate and controversy on normativity started from the first centuries after the foundation of the religion by the Prophet Muhammad. One of the very first questions that occurred to the forefathers of Islamic Theology was on the theological bases of normativity. This debate divided Muslim theologians into two major branches: the *Mu'tazilites,* who believed that independent reason can be relied on as a source of normativity, and the *Asha'rites,* who believed that the only legitimate source for uncovering moral goodness or badness is the Holy Scripture. Shiite theology, after its initial development in the eighth and ninth centuries CE, was more inclined to the *Mu'talizites*' because of the latter's recognition of independent reason as a source of normativity.

Although Islamic theology (*kalam*) provides the theoretical backgrounds of moral thoughts, it is the Islamic Jurisprudence (*fiqh*) that has always been in charge of producing normative guidance and instructions in Muslim communities. Likewise, bioethical issues have been subjected to Islamic Jurisprudence. Most Muslim families ask their bioethical questions to Islamic jurists (*fuqaha*). In addition, in many countries with Muslim majorities, the laws and regulations must be in accordance with *fiqh*. At the same time, the question of normativity has also been a source of controversy among authorities and scholars of *fiqh*. In the field of bioethics, debatable issues such as organ marketing, third-party assisted reproduction, and brain death have made jurists to reconsider their theories on normativity at some points. For example, in the Islamic Republic of Iran, the efforts to make all the health-related regulations compatible with *fiqh*, have raised theoretical debates on the legitimate source of bioethical normativity in Islam and how to deal with the conflicts with the globally dominant Western bioethics and its normativity. Muslim scholars have adopted four main approaches in dealing with such conflicts: (1) The mainstream of scholars and jurists in the major Seminaries believe that *fiqh* is

K. Aramesh (✉)
The James F. Drane Bioethics Institute, College of Sciences and Health Professions, Edinboro University of Pennsylvania, Edinboro, PA, USA
e-mail: karamesh@edinboro.edu

J. Gielen (ed.), *Dealing with Bioethical Issues in a Globalized World*,
Advancing Global Bioethics 14, https://doi.org/10.1007/978-3-030-30432-4_3

a comprehensive source of guidance that is able to answer all the normative questions based on its main traditional sources; (2) The theory of *mantaghat-ul-faragh*; (3) The principle-based approach; and (4) The theory of minimal approach to religion proposed by the leaders of Islamic Intellectual Movement, which has not gained popularity in the mainstream of scholars or believers.

Keywords Islamic bioethics · Islam · Islamic jurisprudence · Shi'a · Sunni · Iranian model of kidney transplantation

3.1 Introduction

In Islam, debate and controversy on normativity started in the very first centuries after the foundation of the religion in the second and third decades of the seventh century CE. One of the very first questions that occurred to the forefathers of Islamic Theology was about the theological bases of moral goodness and badness or, in other words, the theoretical foundations of Islamic normativity. This debate, among others, divided Muslim theologians into two major branches: (1) the *Mu'tazilites*, who believed that independent reason can be relied on as a source of normativity and the Holy Scripture has to be understood and interpreted in the light of independent reason, and (2) *Asha'rites*, who believed that the only legitimate source for discovering moral goodness or badness is the Holy Scripture.

The above division occurred in the Sunni denomination of Islam. The Sunni denomination encompasses 80–90% of the Muslim population. The second main denomination is Shiite Islam, encompassing almost the remaining 10–20%. Shiite theology, after its first development in the seventh and eighth century, was more inclined to the *Mu'tazilites* because of its recognition of independent reason as a source of normativity.

The theoretical roots of the Islamic normativity have indeed to be sought in the Islamic theology (*kalam*). However, the normative system of Islam has been best embodied in the Islamic divine law (*shari'a*). The branch of Islamic knowledge in charge of discovering and presenting of *shari'a* to the believers is Islamic jurisprudence or *fiqh*.

In the Muslim world, *fiqh*, not *kalam*, has always been used to generate normative guidance and practical instructions for the believers. Islamic Bioethics has not been an exception (Shomali 2011; Ghaly 2014). Most Muslim families and healthcare providers ask their bioethical questions (e.g. questions about assisted reproduction, abortion, organ transplantation, euthanasia, etc.) to experts in Islamic jurisprudence (*ulama* or *fuqaha*[1]). In addition, in many countries with Muslim majorities, the laws and regulations must be in accordance with *fiqh* (Larijani and

[1]The word *ulama* is the plural form of *alim* in Arabic. *Alim* means scholar. However, in its usual usage, it refers to a scholar who is an expert in Islamic jurisprudence. *Fuqaha* is the plural form of *faqih*, which means a scholar who is an expert in *fiqh*. The general meaning of *fiqh* is deep knowledge. However, this term is almost always being used as equal to Islamic jurisprudence.

Zahedi 2006; Aramesh 2007). Yet, the question of normativity has also been a source of controversy among authorities and scholars of *fiqh* (Sachedina 2009a, 3–76).

In the field of bioethics, controversial issues such as organ marketing, third-party assisted reproduction, family planning, and distributive justice have made Islamic scholars to reconsider their theories on normativity at some points (Aramesh 2009a). Over the past decades, such bioethical issues have raised some major debates and controversies in the scholarly and judicial circles, mainly because the traditional approaches adopted by Islamic jurists and scholars proved to be partly in conflict with the global common morality. Therefore, the traditional reason vs the Holy Scripture debate reappeared in the new form of contemporary (modern Western) rationality vs traditional understandings of the Holy Scripture. As an example, in the Islamic Republic of Iran, the efforts to make all the health-related regulations compatible with *fiqh*, have raised theoretical debates on the legitimate source of bioethical normativity in Islam and how to deal with the conflicts with the globally dominant Western bioethics and its normativity (Larijani and Zahedi 2006; Larijani et al. 2005; Zahedi and Larijani 2008; Aramesh 2008a).

This chapter takes a closer look at the historical and current debates between reason-based and scripture-based normative schools in Islamic theology. It, then, portrays the current controversies raised in the field of Islamic bioethics as a result of the newly emerged conflicts between traditional *fiqh* and modern rationality in Muslim societies. Next, this chapter will categorize the main current normative approaches in the field of bioethics into four main categories, based on their approach to how to solve the above-mentioned debates and conflicts. These four approached are: the mainstream of *fuqaha* who believe in the self-adequacy of the traditional *fiqh*, the proponents of the theory of *mantaghat-ul-faragh*, the developers of a principles-based approach, and the proponents of the minimal approach to religion. These approaches are described with more details, below in this chapter. Finally, this chapter will portray how these different approaches deal with practical bioethical issues in the real world and depict their similarities and differences by using the issues raised by kidney transplantation as a case study.

It should be mentioned that categorizing the various approaches of the numerous and diverse branches of Islamic thoughts into a limited number of main categories is only achievable at the cost of some degrees of generalization. Consequently, a detailed picture of all diversity may not be obtained. However, this chapter tries to provide the most inclusive model to encompass almost all the existing approaches that are (or can potentially be) influential in bioethical debates in the contemporary Muslim world.

Although all the main Islamic perspectives toward normativity in bioethics are broadly covered, however, for providing practical and real-world examples, this chapter mostly relies on the experience of the Islamic Republic of Iran in founding and developing Islamic approaches to bioethics. This was because of the first-hand experience of the author on the bourgeoning field of bioethics in Iran and because of the unique characteristics of Iran's experience, including its pioneering in many branches of healthcare in the Muslim world, while at the same time trying to regulate its health system and develop biomedical ethics based on Islamic *fiqh* as the foundation of all legislation.

3.2 Methods and Sources of Islamic Normative System

As mentioned above, Islam as a religion has two major branches: *Sunni* and *Shi'a*. The majority of Muslims in the world are *Sunni*. The majority of the population in most Muslim-majority countries, such as Saudi Arabia, Egypt, Indonesia, Malaysia, Pakistan, Turkey, are Sunni. *Shi'a* Islam encompasses 10–20% of all Muslims living as majorities in a few countries, most notably Iran and Iraq, and as minorities in many other Muslim countries and societies. Although at the beginning, a debate on who was the legitimate successor of the Prophet Muhammad started the Sunni-Shi'a division, through the course of the history, these two branches developed other differences with each other in terms of theology, jurisprudence, and normativity. Some aspects of these differences are described below in this chapter because the differences between these two branches of Islam also impact normative attitudes to bioethical issues.

A number of books and numerous articles have been written on the Islamic perspectives towards different bioethical issues. Although informative and clarifying, these essays may hide the diversity that exists among various ethical-legal schools within Islam. Sometimes, especially when it comes to *Shi'a* vs. *Sunni* debates, the differences are huge. In these cases, usually authors depict the perspective of only one sect of Islam as the only Islamic view, while, in reality, this view only belongs to that particular sect. From the viewpoint of people outside the religion, this may lead to one-sided interpretations of Islamic bioethics. For example, while analogical deduction (*qiyas*) is one of the four main sources of jurisprudence in most Sunni schools, it has been rejected and even condemned in Shiite schools and replaced by other forms of reasoning (*aql*) with specific conditions and limitations (Motahari 2001). Such differences lead to very diverse normative outcomes. For example, while most *Shi'a* authorities have permitted surrogate motherhood and embryo donation, most of their *Sunni* counterparts have strictly forbidden the same practices (Aramesh 2009b).

Islamic Jurisprudence (*fiqh*), as an authentic legal system, which delineates divine law (*shari'a*), has enormous influences on every aspect of the lives of Muslims. There are various branches of *fiqh* in both Sunni and Shi'a Islam, but all use four main sources.

1. The Qur'an: The Qur'an is the most important part of the Holy Scripture for all the Muslims. An absolute majority of Muslims believe that Qur'an literally encompasses the words (messages) of Allah directly descended to the Prophet Muhammad and he, as a truthful and infallible messenger, delivered them to the Muslims. Qur'an is the most important and consensual source of normative direction and guidance for all the branches of Islam. There is a huge body of normative guidance and instructions in the Qur'an. Although different in their interpretations, all major branches of Islamic thought regard the Qur'an as the cornerstone and foundation of the Islamic normative system (Sachedina 2009a, 5; Motahari 2001; Mohaghegh Damad 2012; Atighetchi 2007).

2. *Sunnah:* It encompasses what the Prophet Muhammad (in Sunni Islam) or the Prophet Muhammad and the Holy Innocent Imams, (in Shi'a Islam) have said (*hadith*), done, or agreed upon. This part of the Holy Scripture is recorded and preserved in a series of books (the most trusted ones are 4 major books in Shiite and 6 major books in Sunni branches). This is one of the most important differences between the major branches of Islam. While Sunni Muslims regard only the *sunnah* of the Prophet Muhammad as a part of the Holy Scripture, Shiite Muslims believe that Imam Ali bin AbiTalib (the cousin and son-in-law of the Prophet) and a certain group of his descendants (the Holy Innocent Imams) were the legitimate successors of the Prophet Muhammad in his religious authority.[2] *Sunnah* takes the second place after the Qur'an in the degree of legitimacy in shaping normativity in Islam. Some Shiite scholars believe that it is impossible to have a correct understanding from the Qur'an without interpreting it in the light of the *sunnah* of the Prophet Muhammad and the Holy Innocent Imams. Sunni scholars, in contrast, do not regard the Sunnah of the Holy Innocent Imams a necessary element for a correct understanding of the Qur'an and the prophetic *sunnah* (Motahari 2001; Mohaghegh Damad 2012; Aramesh 2008b).
3. *Ijma* or consensus among Islamic scholars: Although this is an important source in Sunni *fiqh*, in Shiite *fiqh*, *ijma* is valid only with endorsement of an innocent Imam (Motahari 2001). Sunni Muslims, who have always had an absolute majority in the Muslim world, regard the community of Muslims and Muslim scholars as a kind of embodiment of the prophet. Therefore, the consensus among Muslim scholars, for them, is equivalent to the sunnah of the prophet. Sunni Islam is also called the religion of *sunnah* and *jama'a* (tradition and community). Shi'a Muslims, on the other hand, have never accepted the consensus of Muslims as a valid source of normativity. This disagreement first came up just after the death of the Prophet Muhammad. When the community of Muslims elected *Abu Bakr* as his successor, a group, who would later be called Shiites, asserted that the successor of the prophet had already been appointed by him and was Ali Bin AbiTalib. Since then, the shi'a Muslims have always believed that the consensus among Muslim is not valid unless it is endorsed by a divine authority, that is the Prophet Muhammad or one of the innocent Imams.
4. The unanimous acceptance of *ijma* has been eroded by a few theories developed by Muslim scholars. For example, Abdulkarim Soroush, an Iranian Muslim scholar believes that, even in the Holy Scripture, there are two parts: the part that is essential to the religion and has to be considered categorical (this part does not include jurisprudential items) and the other part that is accessory and is just determined by the historical, cultural, and societal contexts in which Islam developed and the prophet Muhammad lived. That part, therefore, is subject to change and can be replaced by new understandings and findings of human reason and

[2] Different branches of Shi'a Islam vary in terms of the number of descendants they believe to be Holy Imams. The largest branch, which is dominant in Iran, Iraq, and Lebanon, believes in 12 Innocent Imams. Adherents of this branch believe that the last one of these Imams has disappeared and is still alive and will come back in the future to revitalize Islam in the world.

rationality even if the underlying convictions are included in the Qur'an. This theory has not gained popularity among Muslims and is just discussed and accepted among a small fraction of intellectual Muslims.

5. *Qiyas* in Sunni Islam and *aghl* in Shiite Islam: *Qiyas* means analogical deduction and *aghl* means reason (Motahari 2001, 16–24; Al-Bar and Chamsi-Pasha 2015). Using analogical deduction as a source of *fiqh* means that when the *Sunni* jurists encounter new ethical issues or questions, they search the Holy Scripture to see how it has treated similar issues or questions and try to find an answer based on these ancient solutions. Traditionally, for *Shi'a* scholars, using independent reason means relying on the faculty of reason in a way that was understood by the philosophies rooted in ancient Greek culture (Tusi 2008; Ghoochani 2016; Dabbagh and Aramesh 2009).

The most important ones among these resources are the Holy Quran and *sunnah.* In what follows, we will jointly refer to these two resources as Holy Scripture. The mainstream of Muslim jurists accept that when there is a relevant assertion in either the Holy Quran or *sunnah*, it should be considered categorical and no other argument can overrule such a clear assertion. As mentioned above, in Shiite theology, in contrast with the majority of *Sunni* schools, independent reason is accepted as a source of discovering moral goodness and badness (Mohaghegh Damad 2012). However, there is no similar tendency towards the acceptance of independent reason in the dominant traditional schools of *fiqh* in *Shi'a* Islam.

Reason, however, has not been entirely abandoned. Both Sunni and Shiite scholars have listed a series of aims (*maqasid*) for Islamic law. According to these scholars, Islamic law has been created by God to safeguard, keep, or achieve these aims. Although there have been controversies on the exact number of items on this list, there is general agreement that the list includes preservation of faith (din), life (al-nafs), mind (al-'aql), progeny (al-nasl), and property (al-mal) (Al-Bar and Chamsi-Pasha 2015; Zamani and Qiam-oddin 2014).

Theoretically, specifying a set of aims for the Islamic system of normativity could give it some degrees of secularity, because most of these reasons are similar to the ones that would be set by secular reasoning. Also, having a set of specific aims makes the rules and regulations instrumental in nature. In reality, however, tendencies towards secularism in Islam have been limited. Instead, there has never been an end to the controversies and debates between the proponents of relying solely on the Holy Scripture (*Asha'rites* in Sunni Islam and *Akhbari'a* in Shiite Islam) and the proponents of giving a place to the independent reason (*Mu'tazilites* in Sunni Islam and *Osuli'a* in Shiite Islam). In the modern era, in addition to the previously ongoing debates on the role of the reason, debates on the role of rationality have emerged among Muslim scholars. As mentioned above, such debates came up as a result of the conflicts between the modern and Western rationality and the traditional accounts of *fiqh*. As a part of the answer to these debates, some Muslim scholars have tried to formulate a set of principles for Islamic biomedical ethics (Sachedina 2009a, 25–76). A more detailed discussion on this approach has been provided below in this chapter. As far as it can be related to this part of this chapter, this approach shares the same sources and aims as described for *fiqh* above.

In the following parts of this chapter, these two aspects on the normativity debates in the Muslim world (reason vs. the Holy Scripture and modern rationality vs. the traditional normative bases of Islamic jurisprudence) are discussed in more detail.

3.3 Reason vs. Holy Scripture: A Historical Debate

As per the Qur'an, the human beings are simultaneously vicegerent (*khalifa;* Qur'an, 2:30) and slaves (*Ibād*; Qur'an, 37:40) of God. Either way, the Qur'an asserts that human beings are responsible for their deeds (Qur'an, 74:38). This assertion of responsibility has been called "the essence of morality in the ethics of Islam."(Al-Bar and Chamsi-Pasha 2015, 50) On the other hand, the moral responsibility of the human beings as rooted in their autonomy can be considered as self-evident and approved by independent reason. This shapes the other cornerstone of morality especially in Shi'a Islam (Ghoochani 2016; Tusi 2008).

But, then, if human beings are responsible to act morally and this responsibility is confirmed by both the Holy Scripture and independent reason, how can they discover moral goodness or badness in fulfilling this responsibility? Can they rely on the independent reason (as the legitimate vicegerents of Allah, who has bestowed on them independent reason and autonomous will) or must they just seek what their lord has asked them to do (like what the real slaves do)? As described above in this chapter, this very question divided Muslim scholars into two main theological branches: *Asha'rites* and *Mu'tazilites*.

In the ninth century CE, the Translation Movement dramatically changed the perspective and limits of knowledge, science and wisdom in the Muslim world. The Translation Movement was a collective endeavor by Muslim scholars over the ninth and tenth centuries CE that included the translation and interpretation of a massive number of the sources of Greek philosophy and other branches of knowledge into Arabic. It occurred in prosperous Muslim cities such as Baghdad and Damascus under the ruling of Abbasi Caliphates. In addition to works on Greek philosophy, books in other languages, such as ancient Persian (Pahlawi) and Sanskrit on topics as diverse as medicine, astronomy, mathematics, and ethics were translated into Arabic during the Translation Movement. The effects and implications of this movement for Islamic civilization were tremendous. The subsequent flourishing of almost all branches of human knowledge in the Muslim world was largely owed to the Translation Movement.

It was not surprising that after the Translation Movement occurred, philosophy began to flourish among Muslim scholars and some prominent figures, such as Avicenna (Ibn Sina) (980–1037 CE) and Averroes (Ibn Rushd) (1126–1198 CE) created seminal books in philosophy and philosophical ethics.

This flourishing, however, did not remain unopposed. In the eleventh century CE, another famous Muslim scholar, Abu Hamid Al-Ghazali (1058–1111 CE) argued that the philosophical works of the Muslim philosophers, who were inspired by the Greek philosophy, were in disagreement with the original and authentic Islamic beliefs and norms. He argued that the inspiration and influence of Greek philosophy had distorted Islamic normativity, since, in his view, authentic Islamic normativity is only achievable through studying and exploring Holy Scripture (Al Ghazali 1963). Al Ghazali was followed by other influential Muslim scholars and thinkers such as Rumi (Molana Jalaleddin Balkhi) (1207–1273 CE) whose seminal book in Persian, *Mathnawi*, is still the most important book in Muslim Sufism and Mysticism and has been very influential in shaping different aspects of normativity in the Muslim world (Rumi 1996).

As a matter of fact, the debate between *Ash'aria* and *Mu'tazila* in the Sunni world ended with the dominance of the *Ash'aria*. Therefore, in the mainstream of Sunni jurisprudence, independent reason has no place and is replaced by analogical deduction based on Holy Scripture. It was only in recent decades that a movement aiming at revitalizing the *Mu'tazila* approach appeared on the scene of Sunni schools of Islamic thought (Soroush n.d.).

In the Shiite branch of Islam, however, the course of events was somewhat different. Both Shiite theology and *fiqh* kept recognizing independent reason as a source of normativity (Ghoochani 2016; Mohaghegh Damad 2012). Over the course of its history, the Shiite *Fiqh* was divided into two main branches. A branch, called *Akhbari'a,* was proponent of reliance only on the Holy Scripture for discovering Islamic normativity. This group of jurists believed that the independent reason is valid in identifying the very general norms (e.g. justice is good), but cannot be relied upon in deliberations on more practical rules and regulations. This branch did not achieve a widespread acceptance in Shiite seminaries. The other branch, called *Osuliy'a*, used a system of interpreting the Holy Scripture in the light of reason. The process that is called *ijtihad.* This branch prevailed and achieved dominancy in Shiite seminaries. However, this does not mean that normativity in Shiite jurisprudence is primarily based on independent reason. The mainstream of Shiite *fuqaha* restricted the role of independent reason to a few consensual and general principles[3] and limited the function of reason as the fourth source of *fiqh* to discovering the unexpressed ideas or viewpoints of the Innocent Imams.

In Shiite *fiqh,* the idea of independent reason can be seen in the concept *Bana-e-uqlal.* This concept refers to what is created or agreed upon by reasonable people, regardless of their religion, nationality, or race. It has been accepted as a valid argument in the absence of an assertive guidance in the Holy Scripture. The legitimacy of *Bana-e-uqlal* has been used by a group of Shiite jurists to create an approach

[3]These principles are called *mustaqellat e aqliyyeh,* which means "what reason can discover on its own." There are just a few consensual principles of this kind such as "Justice is good." In reality, these principles are too general to be useful in practical debates. For example, although justice is good, the mainstream of *fuqaha* believes that exact meaning of justice and its implications have to be determined in the light of the Holy Scripture.

different from the mainstream in dealing with newly-emerged ethical issues. However, the concept has not been used by the mainstream of *fuqaha* in a way that it can play a significant role in resolving the cases of conflict between Holy Scripture and independent reason (Hashemi and Torabi 2014).

In the Islamic world, the debate on the role of independent reason has almost always ended with the dominance of reliance on the Holy Scripture. However, in the modern era, this historical approach has encountered new challenges that cannot be solved with the traditional methods and cannot be discussed and understood with the traditional approaches to the conflicts between independent reason and Holy Scripture. These newly-emerged challenges and their implications are discussed in the next part of this chapter.

3.4 Rationality vs. Traditional Jurisprudence: A Modern Debate

The modern era has opened a substantially different new chapter in the debate between the *Ash'arites* and *Mu'tazilites*. Many Muslims, particularly those who have enjoyed good education, have accepted modern rationality along with other aspects of modernity, such as scientific and technological advances, modern educational systems, bureaucracy, and judiciary bodies. As a consequence, in Islamic societies, traditional ethical and legal systems have encountered a serious challenge: they have had to prove their efficiency and adequacy in hard competition with modern and secular systems. Muslim scholars have responded to these challenges in mainly three ways:

1. The first group of Muslim scholars regard the traditional *fiqh* as the only legitimate and rightful source and method of normativity. In the cases of conflict with modern rationality, they simply suggest Muslims to give absolute priority to the traditional *fiqh* rooted in Holy Scripture. According to this group, Islamic *fiqh* is a complete, sacred, and self-sufficient ethical and legal system that does not need to be amended or influenced by any other system of thought.
2. The second group tries to align modern rationality and Islamic normativity. In the field of biomedical ethics, the proponents of the theory of *mantaghat-ul-faragh* and the principle-based approach, which will be both described below, are the most prominent examples of scholars in this group. This group aims at preserving the major elements and cornerstones of the traditional *fiqh* while making it compatible with the established modern ways of life and globally accepted norms.
3. The third group finds the compatibility and reconciliation between the traditional *fiqh* and modern rationality theoretically impossible. They argue that the traditional *fiqh* is not an essential part of Islam and should be replaced with a secular ethical and legal system. According to this group, the essential part of Islam is its mystical message and its teaching about the existence of God and the Day of Judgment.

3.5 Four Approaches to Normativity in Islam

As explained above, one can argue that the main challenge for Muslim jurists and scholars in shaping their framework of normativity is the relation between the Holy Scripture and the independent reason and modern rationality. Accordingly, Muslim scholars can be divided into four main groups regarding their views on how normative arguments are to be constructed in Islam. For portraying a clearer picture, this part of this chapter tries to depict these main groups in contemporary Iran. However, almost the same groups and positions can be found and traced in the Sunni world:

(1) The mainstream of scholars in the seminaries, including the Seminary of Qom, the largest and most important seminary of Shiite Islam, believe that Shiite *fiqh* is a comprehensive source of guidance that is able to answer all the normative questions. In the view of most Shiite scholars, Shiite *fiqh* provides an all-inclusive ethical and legal system. This means that *fiqh* is able to answer all the questions about the goodness or badness of humans' voluntary deeds. According to this perspective, every voluntary act of a human being can be included in one of these *fiqh*-defined categories: Mandatory (*vajib*), Preferred/Recommended (*mustahab*), Allowed (*mubah*), Discouraged/Detested (*makruh*), and Forbidden/Prohibited (*haram*) (Motahari 2001, 66–7). Therefore, no voluntary action of human being falls outside of the coverage of Islamic *fiqh*. Consequently, there is no place for any other system of value judgment including the secular ones, namely applied ethics, in generating the normative system of Shiite Islam. Therefore, according to the mainstream of Shiite religious scholars, Shiite bioethics and medical ethics is identical to Shiite bio- or medical jurisprudence (*fiqh*) (Mohaghegh Damad 2012, 34–6).

(2) The proponents of the theory of *mantaghat-ul-faragh* proposed by Ayatollah Seyyed Mohammad Bagher Al Sadr and other versions of this model. Some versions of this theory have been dominant in the contemporary medical ethics in Iran. According to this approach, beyond the legal and ethical guidance and instructions asserted in the Holy Scripture, there is an area of silence for which no guidance is available in the Holy Scripture. All the issues that fall within this area can be examined and solved in accordance with reason. Different versions of this approach vary in term of the extend of areas that they included within the *mantaghat-ul-faragh*. This approach leaves a large part of contemporary bioethics, such as issues surrounding newly-emerged technologies, beyond the reach of the traditional jurisprudential methods of *fiqh* and allows Muslims to appeal to contemporary rationality in dealing with these issues. This approach may seem to be helpful in solving many of the conflicts, but it suffers from two limitations: First, since the founder of this approach, Ayatollah Seyyed Mohammad Bagher Al Sadr (1935–1980), was executed by Saddam Hussein's regime at a relatively young age, he did not find enough time to deliberate his theory. Second, this theory has not been accepted and adopted by other Muslim jurists. If an ethical or legal question is asked from a Muslim authority, he will never reply that this is related to the area of silence and the question has to be

referred to an ethicist. In contrast, believing in the inclusiveness of *fiqh*, Muslim authorities consider themselves qualified to answer every ethical and legal question. Nevertheless, as we will see later in this chapter, the theory of *mantaghat-ul-faragh* is unconsciously applied in practice. For example, in Iran, ethicists, policy makers, and healthcare workers prepare the draft version of ethical guidelines and the religious authorities just review them to verify that they do not go against the assertions of the Holy Scripture (Aramesh 2014b; Aramesh 2009a; Zahedi and Larijani 2008; Larijani et al. 2006).

(3) A group of Muslim scholars have proposed a principle-based approach. Among them, Abdulaziz Sachedina in his book titled *Islamic Biomedical Ethics* tries to develop an Islamic framework of bioethical normativity by providing a set of principles (Sachedina 2009a, 25–76). It is important, however, to note that the principle-based approach in Shiite jurisprudence differs in three important ways from the principle-based approaches in the mainstream Western bioethics, like the four-principle approach formulated by Beauchamp and Childress (2013).

First, the four-principle approach is – or was intended to be – a comprehensive ethical framework for biomedical ethics, while the principles of Shiite jurisprudence do not consist of a comprehensive system for ethical-legal reasoning and deliberation. In other words, these principles are used in just a part of such deliberations, along with other –sometimes more important- parts such as the assertions of the Holy Scripture or the consensus of jurists. In a large part of the scholarly deliberations, various parts of the four sources, like the verses of holy Quran or *sunnah*, are used without appealing to any principles.

Second, the four-principle approach is based on secular and independent reasoning while the principles of Shiite jurisprudence are not based on independent reasoning. These principles are derived from the Qur'an or Hadith or other parts of *sunnah* to help jurists to extract practical decrees from the Holy Scripture (Motahari 2001, 30–55). For example, the principle of *la-haraj* (not imposing unbearable responsibilities or other difficulties on people) is derived from some verses of Qur'an like the one in the surah of al-haj which says: "... He has not placed upon you in the religion any unbearable difficulty..." (Qur'an 22:78).

Third, the four-principle approach encompasses a set of principles that each of them provide a general guidance for a vast variety of practical issues, while the principles of Shiite jurisprudence are not always broad, but sometimes they are narrow in the extent of the practices they guide. For example, a principle called *ferash*, holds that a child born to a household should be considered the legitimate offspring of the man who is the head of that household. This principle cannot be compared with general principles such as beneficence or justice.

(4) Some reformist thinkers have recently invoked the Minimal Approach to Religion, recognizing independent reason in Shiite theology to suggest fundamental reforms in the current *fiqh* making it more compatible with the contem-

porary universal norms of human rights and freedoms (Soroush n.d.). By Minimal Approach to Religion, they mean considering religion in charge of a minimal set of functions such as guiding the believer about the afterlife, and not expecting other functions such as legislation or political leadership from the religion. This group of thinkers, also called the intellectual movement, has not gained much popularity and acceptance among the mainstream of scholars or believers. Therefore, in studying Islamic bioethics, most authors usually ignore this group and just review the perspectives of the traditional, conservative mainstream of Shiite *fiqh*.

3.6 A Case Study: The Iranian Model of Kidney Transplantation

In order to illustrate how the above-mentioned approaches to normativity in Islamic bioethics concretely operate, we will examine the different Islamic responses to the ethical issues raised by live-donor kidney transplantation in Iran.

After the foundation of the Islamic republic of Iran in 1979, the relation between *fiqh* and law/ethics became a prominent issue in that country. According to the constitution of Iran, all laws and regulations must be rooted in and in accordance with Islamic (Shiite) *fiqh*. For this purpose, in addition to involvement of Islamic scholars and authorities in compiling and drafting legislations, the Guardian Council reviews all the legislations passed by the Islamic Parliament to ensure the above-mentioned concordance. From time to time, some conflicts arise between common morality and Shiite *fiqh*. Common morality, or the core and basic moral beliefs that are common among the educated healthcare workers, is highly influenced by the contemporary Western rationality. Therefore, the abovementioned conflicts with the traditional jurisprudence frequently happen.

In some cases, such conflicts with common morality and/or human rights result in difficulties for the Islamic republic at both the domestic and international level (Sachedina 2009b). To resolve such problems, Ayatollah Khomeini, the founder and first supreme leader of the Islamic Republic of Iran, established a new institution: The Expediency Discernment Council of The Islamic Republic of Iran. The role and functioning of this institution is described in the revised Iranian constitution of 1988. The major function of this council is reconciliation between the parliament and the Guardian council in the cases in which the parliament ratifies an act and insists on its necessity and importance for the country but the *fuqaha* in the Guardian Council reject that act because of its contrast with the teachings of *fiqh*. An examination of the extent to which this council has been successful in resolving conflicts between *fiqh* and common morality is beyond the scope of this chapter.

In Iran, the first renal transplantation was carried out in 1967 in Shiraz, a city in the southern part of Iran (Larijani 2010). By 1978, the year of the Islamic revolution, a total of about 100 surgical renal transplantations had been carried out. In many cases, kidneys used for transplantation were obtained and imported from the European Transplantation Network (Larijani 2010; Mahdavi-Mazdeh 2012). After the revolution and subsequent foundation of the Islamic Republic, the renal transplantation programs were stopped for a few years. In the early 1980s, patients who needed a renal transplant had to seek this treatment abroad, mostly in the United Kingdom. This option was unaffordable for most of the patients. At the same time, because of the shortage of facilities needed for hemodialysis, particularly during the Iraq-Iran war (1980–1988) a large number of such patients died (Larijani 2010). During the second half of the 1980s, a number of Iranian surgeons completed training in kidney transplantation in Western countries. When they returned to Iran, they formed renal transplantation programs in University hospitals. Consequently, numerous renal transplantations were performed using living organ donors (Larijani 2010).

Before establishing these programs, however, the surgeons needed the permission of the religious authorities. The health policy makers, the physicians, and their patients had two main questions that these authorities needed to answer:

1. Is transplanting an organ from one person to another allowed by Islamic *fiqh*?
2. If so, under what conditions and circumstances?

The answer to the first question was very simple and straightforward. If a transplant is necessary for the health of the recipient and is not fatal or unbearable for the donor, this is allowed. Therefore, collaboration among physicians and religious authorities (in the form of providing information by the trusted physician and issuing *fatwas* by the authorities aimed to solving the problems encountered by Muslim physicians and health policy-makers) paved the way for formal development of the Iranian organ transplantation policy that was enacted in 1988 (Larijani et al. 2004; Aramesh 2009a). That policy, however, was shaped by the specifically Iranian context. As we already mentioned above, in the 1980s, a large number of patients with end stage renal disease needed renal transplants and made up a long waiting list of people waiting to travel abroad for kidney transplantations, using governmental support. The financial burden, along with the deficiency of dialysis facilities in the 1980s, pushed health authorities to find a way out. They designed and proposed a model based on a regulated market for live kidney transplantation.

Now, it was the time for the religious authorities to deal with the second question as listed above. As a matter of fact, most Shiite authorities do not regard the selling of organ parts right for Muslims because the religious scholars do not regard human body and its parts as commodities. However, in order to obliterate the kidney waiting list, they permitted monetary compensation for kidney donation. The religious decrees (Fatwas) of religious authorities prepared for the establishment of a compensated, living unrelated donor renal transplantation program, which was named the Iranian model of kidney transplantation (IMKT). After establishment of this model, the number of renal transplantations that were performed in Iran increased speedily and, by 1999, the renal transplant waiting list was eliminated (Ghods and Savaj 2006). Reportedly, Iran has the largest program of unrelated live donor kidney transplantation in the region (Einollahi 2010; Simforoosh 2007). This model, however, has always been a subject of ethical review and criticism in Iran (Bagheri 2006; Aramesh 2014a, 2016). In responding to these ethical problems, theoretically, three main approaches could have shaped the Iranian organ transplantation policy.

1. In order to assess the IMKT, legislators and theologians could have relied on the approach of the mainstream of the Shiite seminaries, which implies relying on an evaluation by the *fuqaha*. In practice, this approach proved to be impossible because of two main reasons. First, there is not enough expertise in the Shiite seminaries to deal with such technical medical issues in depth and issue specific decrees or create specific and specialized guidelines. As a matter of fact, most of the religious decrees are issued in response to the questions (*istifta'*) made by the stakeholders, e.g. practicing physicians and policy makers. In many cases, the questioners design their questions in a way to obtain their desired responses e.g. religious permission of a specific kind of practice. This has led to a group of *fatwas* that are incompatible with the contemporary norms of practice or resulted in conflicts and problems in practice. Other examples include surrogate motherhood (Aramesh 2009b) and sperm and embryo donation (Afshar and Bagheri 2013). In the latter case, paying no attention to the consequences of the current practices (e.g. anonymity of donors) resulted in ratification of a very flawed law. Second, in some cases, the opinions of *fuqaha* are in conflict with the contemporary norms and standards of practice that are learned and believed by healthcare professionals. For example, giving priority to Muslims on the waiting list for organ donation (which is recommended by the traditional understandings of *fiqh*) cannot be included in the guidelines, nor can healthcare professionals be expected to observe such instructions.
2. Because of the above problems, the reality of the health sector in Iran is that the policy-makers follow the second approach, which is based on the theory of *mantaghat-ul-faragh*, in creating the ethical guidelines and regulations on the details and technical specifics of practices such as organ transplantation. This means that the main body of these regulations and guidelines are created based on the global experience and standards and then endorsed by religious scholars (Aramesh 2007, 2009a; Zahedi and Larijani 2008).
3. The other approaches described above, including the principle-based approach and the one that belongs to the intellectual movement, have not been used in the Iranian debates on organ donation. These approaches have not become popular enough, either among policy makers or healthcare professionals and patients, and have not been adopted by enough bioethicists to create an influential body of directives, guidelines, or dedicated publications. However, theoretically, one can argue that the principles-based approach would be similar to the mainstream since it would consider just the principles accepted by Shiite authorities. Also, according to the general approach of the main figures of the intellectual movement, one can say that in the practical cases such as organ transplantation, the proponents of the intellectual movement do not rely on the specific recourses and

methods of *fiqh*. Instead, they appeal to the general principles and values derived from Islamic thoughts, such as human dignity and the sanctity of human life. Therefore, their approach will be very similar or almost identical with the secular approaches. In other words, the proponents of the intellectual movement give priority to the secular rationality over the traditional understanding of the Holy Scripture (Soroush 1994, 1995).

3.7 Conclusion

The debate on the methods and sources of normativity in Islam is ongoing. As depicted in this chapter, the debates on the role of the independent reason and modern rationality have shaped a large part of the history of Muslim ethics and these issues are still the major sources of controversy in Islamic biomedical ethics. Different approaches have been developed in the course of history to deal with theoretical and practical issues in the field of Islamic ethics and bioethics. The case of the Iranian Model of kidney Transplantation shows that using *fiqh* as basis for developing legislation in modern states has limitations, but so do other Islamic approaches. It seems that none of the above approaches is bestowed with absolute superiority to the other ones in terms of providing a sound and consistent normative framework. Rather, each of them has noticeable flaws. However, almost all of them are still alive and play their own role in Islamic bioethics. The reality of Islamic normativity is the result of the interplay and interactions among all these major approaches.

Acknowledgement Part of this study has been done during my stay at the Brocher Foundation in Geneva, Switzerland in 2009. I would like to express my many thanks to the Brocher Foundation for providing me with an ideal and superb place for studying, thinking, and writing during my stay. In addition, I would like to thank my colleagues at the Medical Ethics and History of Medicine Research Center of Tehran University of Medical Sciences, especially Dr. Alireza Parsapour and Dr. Abdussalih Ja'fari Kermanshah for their great and helpful comments. In addition, I am so grateful to Dr. Joris Gielen, for inviting me to contribute in this volume and for his so helpful comments and suggestions on my manuscript.

Suggested Readings

Among the following publications, the first one provides a more detailed picture on the way that normativity takes shape and becomes implemented in the laws and regulations in Iran. The second one provides a detailed picture of the content of Islamic biomedical ethics and is very helpful in understanding the principle-based approach. The third one also provides a very informative description of the challenges in front of Muslim jurists in the field of biomedical ethics and the ways they deal with these challenges.

Aramesh, Kiarash. 2007. The Influences of Bioethics and Islamic Jurisprudence on Policy-Making in Iran. *The American Journal of Bioethics* 7 (10): 42–44.
Atighetchi, Dariush. 2007. *Islamic Bioethics: Problems and Perspectives*. Dordrecht: Springer.
Sachedina, Abdulaziuz. 2009a. *Islamic Biomedical Ethics*. New York: Oxford University Press.

References

Afshar, Leila, and Alireza Bagheri. 2013. Embryo Donation in Iran: An Ethical Review. *Developing World Bioethics* 13 (3): 119–124.

Al Ghazali, Imam Abu Hamid Muhammad. 1963. *Tahafut Al-Falasifah (Incogerence of the philosophers)*. Trans. Sabih Ahmad Kamali. Lahore: Pakistan Philosophical Congress.

Al-Bar, Mohammed Ali, and Hassan Chamsi-Pasha. 2015. *Contemporary Bioethics: Islamic Perspective*. Cham: Springer.

Aramesh, Kiarash. 2007. The Influences of Bioethics and Islamic Jurisprudence on Policy-Making in Iran. *The American Journal of Bioethics* 7 (10): 42–44.

———. 2008a. Cultural Diversity and Bioethics. *Iranian Journal of Public Health* 37 (1 Sup): 28–30.

———. 2008b. Justice as a Principle of Islamic Bioethics. *The American Journal of Bioethics* 8 (10): 26–27.

———. 2009a. Iran's Experience on Religious Bioethics: An Overview. *Asian Bioethics Review* 1 (4): 318–328.

———. 2009b. Iran's Experience with Surrogate Motherhood: An Islamic View and Ethical Concerns. *Journal of Medical Ethics* 35 (5): 320–322. https://doi.org/10.1136/jme.2008.027763.

———. 2014a. A Closer Look at the Iranian Model of Kidney Transplantation. *The American Journal of Bioethics* 14 (10): 35–37.

———. 2014b. A Brief History of Biomedical Research Ethics in Iran: Conflict of Paradigms. *Developing World Bioethics*. https://doi.org/10.1111/dewb.12053.

———. 2016. Iran's Experience on Living and Brain-Dead Organ Donation: A Critical Review. In *Organ Transplantation in Times of Donor Shortage: Challenges and Solutions*, ed. Ralf J. Jox, Galia Assadi, and Georg Marckmann, 285–292. Cham: Springer.

Atighetchi, Dariush. 2007. *Islamic Bioethics: Problems and Perspectives*. Dordrecht: Springer.

Bagheri, Alireza. 2006. Compensated Kidney Donation: An Ethical Review of the Iranian Model. *Kennedy Institute of Ethics Journal* 16 (3): 269–282.

Beauchamp, Tom L., and James F. Childress. 2013. *Principles of Biomedical Ethics*. 7th ed. New York: Oxford University Press.

Dabbagh, S., and K. Aramesh. 2009. The Compatibility Between Shiite and Kantian Approach to Passive Voluntary Euthanasia. *Journal of Medical Ethics and History of Medicine* 2: 21.

Einollahi, B. 2010. Kidney transplantation in Iran. *Iranian Journal of Medical Sciences* 35 (1): 1–8.

Ghaly, Mohammed. 2014. Islamic Bioethics: The Inevitable Interplay of 'Texts' and 'Contexts'. *Bioethics* 28 (2): ii–iv.

Ghods, A.J., and S. Savaj. 2006. Iranian Model of Paid and Regulated Living-Unrelated Kidney Donation. *Clinical Journal of the American Society of Nephrology* 1 (1136): 45.

Ghoochani, Muhammad. 2016. Khajeh Against Khalifa. *Siyasat-Nameh* 1 (4&5): 7–18 [In Persian].

Hashemi, Seyyed Reza, and Akbar Torabi. 2014. Bana-e-Uqala. http://rch.ac.ir/article/Details/11609. Accessed 25 Nov 2016 [In Persian].

Larijani, Bagher. 2010. *Bioethics in Organ Transplantation: An Islamic Perspective*. Tehran: Medical Ethics and History of Medicine Research Center.

Larijani, Bagher, and Farzaneh Zahedi. 2006. Health Promotion, Islamic Ethics and Law in Iran. *DARU Journal of Pharmaceutical Sciences* 14 (Suppl. 1): 7–9.

Larijani, B., F. Zahedi, and E. Taheri. 2004. Ethical and Legal Aspects of Organ Transplantation in Iran. *Transplantation Proceedings* 36: 1241–1244.

Larijani, Bagher, Farzaneh Zahedi, and Hossein Malek Afzali. 2005. Medical Ethics in the Islamic Republic of Iran. *Eastern Mediterranean Health Journal* 11 (5-6): 1061–1072.

Larijani, Bagher, Hossein Malek-Afzali, Farzaneh Zahedi, and Elaheh Motevaseli. 2006. Strengthening Medical Ethics by Strategic Plannig in the Islamic Republic of Iran. *Developing World Bioethics* 6 (2): 106–110. https://doi.org/10.1111/j.1471-8847.2006.00145.x.

Mahdavi-Mazdeh, Mitra. 2012. The Iranian Model of Living Renal Transplantation. *Kidney International* 82: 627–634.

Mohaghegh Damad, Seyyed Mostafa. 2012. *Medical Fiqh*. Tehran: Hoghughi [In Persian].

Motahari, Morteza. 2001. *Osul e Fiqh-Fiqh*. Tehran: Sadra [In Persian].

Rumi, Jalāl ad-Dīn Muhammad Balkhī. 1996. *Mathnawi*. Vol. 2, 1. Tehran: Elmi-Farhangi [In Persian].

Sachedina, Abdulaziuz. 2009a. *Islamic Biomedical Ethics*. New York: Oxford University Press.

Sachedina, Abdulaziz. 2009b. *Islam and the Challenge of Human Rights*. New York: Oxford University Press.

Shomali, Mohammad Ali. 2011. Islamic Bioethics: A General Scheme. In *Life*, ed. Mohammad Ali Shomali, 15–46. London: Institute of Islamic Studies, Islamic Center of England.

Simforoosh, N. 2007. Kidney Donation and Rewarded Gifting: An Iranian Model. *Nature Clinical Practice Urology* 4 (8): 292–293.

Soroush, Abdolkarim. n.d. I am a Neo-Mu'tazilite. Accessed 4 Oct 2015. http://www.drsoroush.com/English/Interviews/E-INT-Neo-Mutazilite_July2008.html.

———. 1994. *Farbeh-tar az eideolozhi*. Tehran: Serat [In Persian].

———. 1995. *Qabz va bast-e teorik-e shari'at*. Tehran: Serat [In Persian].

Tusi, Khajeh Nassiruddin. 2008. *Akhlagh e Naseri*. Tehran: Kharazmi [In Persian].

Zahedi, F., and B. Larijani. 2008. National Bioethical Legislation and Guidelines for Biomedical Research in the Islamic Republic of Iran. *Bulletin of the World Health Organization* 86 (8): 630–634.

Zamani, Muhammad Hassan, and Qamar-oddin Qiam-oddin. 2014. Comparative Study of the Aims of Shari'a. *Habl-olMatin* 3 (6): 30–49 [In Persian].

Chapter 4
Advancing Death? Discourse and Authority in Judaic Bioethics

Jonathan K. Crane

Abstract This chapter investigates two arenas in which normativity is both acknowledged and contested within Jewish bioethics. The first is a discursive arena that for the most part involves Jewish bioethicists. I consider how these scholars, clergy and professionals discuss dying and death, and identify at least three layers of problematic conversational issues. The first layer is terminological: strong disagreements exist regarding the very definitions of who and what is in the throes of dying. The second layer pertains to intervening in someone's dying. This is complicated by how bioethicists read classic texts. The final layer raises meta-ethical issues of conversing across foundational worldviews. For example, some bioethicists consider moral conundrums predominantly through legal lenses whereas others turn more to narratives and some to theology. It is unclear whether and how such foundations facilitate a coherent Jewish bioethical discourse on, say, death and dying issues. The second arena focuses on the dynamics of authority between rabbis and Jews, especially in regard to thinking about and making decisions regarding dying. This dynamic is most conspicuous in the various advance directives published by the streams of modern Jewry. At one extreme are Orthodox advance directives that mandate signatories to abide by strict interpretations of Jewish law, whereas at the other extreme are Reform advance directives that hardly speak of Jewish law or values, and Conservative forms give options based on positions taken by the movement's centralized committee on law. Examination of such advance directives reveals explicit and implicit notions of rightful authority in regard to making decisions about dying and, ultimately, death. Speaking about and authorizing dying thus illustrate some of the concerns regarding normativity in contemporary Jewish bioethics.

J. K. Crane (✉)
Raymond F. Schinazi Scholar in Bioethics and Jewish Thought at the Center for Ethics, Emory University, Atlanta, Georgia

School of Medicine, Emory University, Atlanta, Georgia

Laney Graduate School, Emory University, Atlanta, Georgia
e-mail: jonathan.k.crane@emory.edu

J. Gielen (ed.), *Dealing with Bioethical Issues in a Globalized World*, Advancing Global Bioethics 14, https://doi.org/10.1007/978-3-030-30432-4_4

Keywords Judaism · Euthanasia · Narrative · Law · Advance directives · Death and dying · Bioethics · Goses · Terefah · Authority

4.1 Introduction

An ancient debate between competing schools is perhaps most apt for this conversation about normativity in modern Jewish bioethics. 2100 years ago in Jerusalem, Hillel the Elder frequently had heated debates with his friend and colleague, Shammai. Indeed, their respective schools of disciples often participated in these extended exchanges. Their topics ranged broadly, addressing all sorts of legal and practical issues. In one instance, their debate lasted for 3 years, with each side insisting that the halakhah, Jewish law, should accord with its position. Ultimately, a *bat kol*, or heavenly voice, interrupted their repartee with the pronouncement that *eilu v'eilu divrei Elohim ḥayim v'halakhah k'beit hillel*—these and these are the words of the living God, and the law accords to the House of Hillel.[1]

Until the interruption, like a qubit in a quantum computer, the norm was in a superposition of both positions. Both sides held valid and reasonable positions, and while the argument lasted, the norm was both. Though this interpretation may be attractive for pluralists, it does not embrace the fact that the interruption finally spun the norm to one side.

One could derive from this Talmudic story that when there is a normative dispute between legitimate authorities, one should wait until a heavenly voice breaks forth from the clouds and declare one side the winner. Another lesson could be that when arguing about a norm, just be sure to be in the House of Hillel. A third lesson could be to avoid getting into such heated debates in the first place because they are interminably long and life is too short for such indecision.

Or, perhaps, one should continue reading the Talmudic story. The Talmud immediately explains why the *bat kol* declared the law according to the House of Hillel. It ruled this way because the House of Hillel was kind and modest and studied the position of the House of Shammai as well as the relevant actions of the House of Shammai before articulating its own opinion on an issue. For such generosity of spirit, the law accords to the House of Hillel. The reason why one position and not the other became the norm was not because it was the more logical, or efficient, cost effective, environmentally sensitive, or internally consistent, nor was it the more holy or theologically sound or righteous position. No – it was made the norm because of the relative humility of its promulgator. The moral stature of its speaker thus serves as the ultimate motivation to consider one position—among other equally valid options—as the rightful norm on a particular issue.

One can therefore derive from this Talmudic story that to ensure that one's normative position become the communally accepted norm, be kind and humble.

[1] Babylonian Talmud (BT) *Eruvin* 13b.

Such calls to be nice are no doubt attractive in this age of snarkiness and skepticism. Indeed, many contemporary disputes would benefit from greater collegiality and patience. Concern about tone and tenor, important as they are, are nonetheless only one kind of concern about the construction of norms in modern Jewish bioethics. Two other arenas also complicate the pursuit of normativity. The first is a discursive arena that for the most part involves Jewish bioethicists. Jewish scholars, clergy and professionals discussing dying and death, for example, face three layers of problematic conversational issues: one pertains to terminology, another to hermeneutics, and the last to meta-ethical commitments. The second arena focuses on the dynamics of authority between rabbis and Jews, especially in regard to thinking and making decisions about dying, and this is most evident in the various advance directives published or promoted by the streams of modern Jewry. These two arenas regarding speaking about and authorizing dying illustrate some of the discursive and pragmatic concerns when constructing norms in contemporary Jewish bioethics.

4.2 Structural Problems

A brief and imperfect review of modern Judaism will contextualize the complexities of both the conversational issues and the movement directives discussed below. Today's Judaism is both like and unlike Judaism of the time of Hillel and Shammai. The advent of the historicizing and critical study of religion in the nineteenth century sparked a cleavage in modern Judaism. Up to that time Jews predominantly shared rites and realities; they were for the most part relegated to second-class status no matter where they lived and because of such social and political isolation their liturgies and rituals were for the most part similar across their communities. The new studies, however, highlighted both change (rabbinic Judaism was newer than and different from biblical Judaism) and continuity (texts, laws and celebrations remained central to modern Jewish life).

Insofar as the new scholarly attitude toward religion championed reason, some Jews found the nonrational features of rabbinic Judaism too far-fetched to maintain. They began to jettison some aspects of Jewish life, like kashrut (rules regulating food development, production, manufacturing, cooking and consumption) and gender segregation; they altered liturgy to be more reasonable, removing references to resurrection of the dead, for example; they thought individual knowledge instead of ancient law should be the arbiter of contemporary behavior. Such Nineteenth Century reforms in Germany were taken across the Atlantic to the United States and became the largest stream of modern Jewry: Reform Judaism. Other Jews were repulsed by these very changes. Wanting to resist such modernization, they retrenched in ritual and rule. They became what is now known as Orthodox Judaism, with a substantial community in the United States, and close-knit ones in Europe and elsewhere. Emerging between these two extremes was an American-born movement: Conservative Judaism. Conservative Jews understood themselves beholden to

Jewish law like their Orthodox neighbors, and they also wanted to benefit from modernity as did their Reform neighbors.

This tri-partite environment facilitated the emergence of distinct yet not radically different ways of creating norms. For Orthodox Jewry, norms are to be discovered within the textual tradition. Precedents exist and even if explicit ones do not, careful analogy can apply old laws to modern conundrums. According to this worldview, the rightful authority figure is the rabbi, insofar as the rabbinate is the institution that put the Jewish legal structure in its place initially and developed it through the millennia. Modern Orthodox Jews, for the most part, defer to their rabbis for norm guidance if not decisions.

Reform Judaism also considers the textual tradition important, but it does not accord it definitive or ultimate authority. Nor does Reform derive its norms primarily from the legal textual tradition. Rather, the contemporary individual is understood as the final arbiter regarding a pragmatic moral decision. It is ideal that that individual be educated in the wealth of wisdom that Judaism has to offer, of course. So it is not uncommon for Reform Jewish institutions to invest in ongoing education, so that its members can make Jewishly-informed moral decisions.

Conservative Judaism, unsurprisingly, plumbs the legal textual tradition as do the Orthodoxy and yet, like the Reform movement, situates those teachings in their historical contexts. In this way Jewish laws and norms are seen as evolving, adjusting to the vagaries of history and science. Modern Conservative Judaism thus has a Committee on Jewish Law and Standards that publishes *responsa*, rigorously researched issues of Jewish law that are nonetheless situated in the contemporary context. While change and adjustment is not abjured, it is neither repulsed altogether.

Additional differences exist among modern Jewish ethicists who work to clarify Jewish perspectives on pressing moral issues of the day. Rhetorically, formalists argue predominantly within a legal paradigm. For them, the law is the primary if not exclusive source for all norms and jurisprudence provides ways to reason about morally complex issues. Narrativists, by contrast, see laws as but one genre among others that exist in the Judaic textual tradition that provide normative guidance. Stories, whether they are biblical or rabbinic in provenance, also offer rich ways of thinking about moral issues, and as such, narratives can be sources of and for norms. A third camp turns more to theological constructs, like covenant, to shape, inform and constrain moral reasoning. Theological concepts can be found in an even broader range of sources: law, lore, even liturgy.

Contemporary bioethicists span the streams of modern Jewry and some do not "affiliate" or identify with any stream. While there is a strong correlation between, say, Orthodoxy and legal formalism, it would be wrong to say that all bioethicists in any stream argue the same way, or that those who argue the same way belong to the same stream. Indeed, cacophony rather than symphony is the norm in contemporary Jewish bioethics.

4.3 Conversing Problems

Given this context, when Jewish bioethicists talk about death and dying, they will invariably confront three difficulties.[2] At the most granular level, disagreement arises regarding central concepts and, specifically, a word central to end of life Jewish discussions. Most contributors to this discourse invoke the word *goses* – a term typically applied to a moribund individual. Many bioethicists consider this term uncontroversial; its definition is what they understand it to be and, because of this fixity, whenever or however it is used it is unambiguous. Such certainty about the meaning of *goses* may be admirable yet it is possible to be inaccurate or outright wrong despite one's certainty.

In recent decades the word *goses* has been understood among Jewish bioethicists according to how J. David Bleich defined it in the 1970s. Bleich, one of the most outspoken Orthodox Jewish bioethicists alive today, insists that a *goses* is an individual who, regardless of interventions, cannot remain alive for longer than 3 days. He emphatically concludes that "Halakhah assumes axiomatically that the death process or the 'act of dying' cannot be longer than seventy-two hours in duration."[3] Since he duplicates this definition many times in his writings, it gets picked up by other bioethicists who defer to his robust scholarship, stature and legal acumen. But a question arises—or should arise—when many defer to a particular authority, modern or classic: though many agree with someone's definition, does this automatically or necessarily mean that that definition is accurate and/or exhaustive?

In which ways is the impulse to defer to another's definition problematic? First, consensus is not the same as correctness. A crowd, while unified, may be misled to believe something to be the case when, in fact, it is not. Second, specifically in regard to the word *goses* and its application to end of life deliberations, technology troubles its temporal limit. Modern medicinal interventions can prolong the inevitable dying process far beyond the otherwise anticipated 3 days had no interventions been implemented in the first place. Moreover, some interventions can keep dead individuals animated long after they have met the criteria for, say, brain death, further blurring the line between the living and the dead. Given such technological prowess, the temporal dimension of the definition of the *goses* seems antiquated.

Another troubling feature of this definition of the *goses* is its heavy reliance upon professional medical judgment. In order to proclaim that an individual has just 72 h left of typical life, experienced medical personnel draw on their direct assessments as well as their reasoned judgment. While guided by statistics, such medical judgment is also grounded upon intuition, hunch and even the practitioner's mood; the profession is called *ars medicina*, the medical arts, for good reason. Medical

[2] This section draws heavily from my unpublished paper, "Who's Dying? Which Death? Contested Issues in Jewish End-of-Life Deliberations," delivered at the Judaism, Science, and Medicine Group conference at Arizona State University, February 2016.

[3] J. David Bleich, "The Obligation to Heal in Jewish Tradition." Published in Bleich and Fred Rosner's *Jewish Bioethics* (New York: Sanhedrin Press, 1979), 34.

judgment is thus not impervious to imperfections, indeed, it can be outright wrong. This means, then, that an individual declared by a doctor to be a *goses* could be wrongly situated in this category. The ramifications of this classification could be dire, for there are rules and regulations, permissions and prohibitions about what can and cannot be done to and for a *goses*. This danger sits, of course, on the outcome side of applying the word *goses*. On the input side of that term, it means Bleich's so-called axiomatic law in fact relies, in substantial part, upon personal subjectivity.

A third problem with Bleich's definition concerns its comportment with the evidence he adduces to it. In his probing essay on this topic, the Talmudic scholar Jeffrey Rubenstein plumbs the classic Judaic textual tradition's development of the term *goses*.[4] His expansive research uncovers various ways this word has been defined and deployed in the very literature Bleich theoretically and probably perused. Though many classic sources admit that a *goses* will die within 3 days, some also hold that a *goses* can linger longer, well beyond 3 days, and, moreover, a *goses* may even recover to live a normal life span. That is, Jewish classic legal sources maintain a more flexible connotation to the category of the *goses* than Bleich acknowledges. Rubenstein demonstrates that Bleich curates his sources to such a degree that he elides the classic debates and controversies about this word and category, and that Bleich also invokes non-legal sources to support his definition. In short, the law and definition of the *goses* is neither as axiomatic nor as narrow as Bleich claims it to be. This throws into question the practice of deferring to a modern bioethical authority who himself defers to minority positions and obscure sources, especially when it comes to something so serious as defining who is imminently dying.

Defining a *goses* along temporal lines as Bleich does is not, of course, the only way to conceive it. Elliot Dorff, another preeminent bioethicist, this time in the Conservative movement, holds that the term *goses* should "include all those who suffer from an incurable, terminal illness, even if it will be a year or more before the person dies."[5] In this way Dorff eliminates the temporal dimension of the *goses* in favor of a diagnosis, specifically the diagnosis of incurability. He goes on to suggest that it would be better to use a different term for this condition, the term *terefah*, or imperiled life, which in his view means someone diagnosed with a terminal, incurable illness. But then, what's the difference between a *goses* and a *terefah*? On his account, Dorff claims that a *goses* means someone who is moribund. While perhaps this is useful, Dorff does not clarify what precisely moribund means.[6] The nominal distinction between *terefah* and *goses* appears to exist without any meaningful difference in substance.

---

[4] Jeffrey Rubenstein, "Can a *goses* survive for more than three days? The history and definition of the *goses*." *Journal of Jewish Ethics*. 2.2(2016), 1–37.

[5] Elliot N. Dorff, *Matters of Life and Death* (Philadelphia: Jewish Publication Society, 1998), 199–200.

[6] Dorff, *Matters of Life and Death*, 232.

What all this means is that there are at least two operating definitions of *goses* in contemporary Jewish bioethical discourse. One understands it along temporal lines and narrow ones at that, and the other considers it according to biological or pathological terms. Such terminological discrepancy critically undermines the coherence of the conversation among Jewish bioethicists, especially when they invoke *goses* in reference to the nearly dead.

On the other hand, both definitions agree indeed that the *goses* is nearly dead. More precisely, the nearly dead is someone else. The *goses* is not a care provider, a family member, a chaplain or rabbi or someone otherwise comparatively healthy. It is someone others declare as such. Coupling the fact that those who are so categorized are vulnerable, exposed and subject to how others view and treat them, with the fact of the ambiguity inhering to the very term *goses* itself together render that concept's use ethically troubling.

This leads to the second layer of discursive troubles, the layer of hermeneutics. Just as dying bodies are variously read, interpreted and labeled, so too are classic religious texts. How corporal and textual resources are read produces profound impacts on how the dying are viewed and treated.

Modern euthanasia conversations frequently assume a conceptual matrix divided by two axes. One axis slides between voluntary and involuntary, the other between active and passive. The quadrants therefore identify active-voluntary euthanasia, passive-voluntary euthanasia, active-involuntary euthanasia and passive-involuntary euthanasia. According to Byron Sherwin, another Conservative bioethicist, active euthanasia includes activities that hasten demise, whereas withdrawal or withholding life-sustaining support constitutes passive euthanasia. Voluntary euthanasia pertains to the dying individual taking action that brings about his or her own demise, whereas action ending a patient's life "without his or her explicit consent" would be involuntary euthanasia.[7]

Of course such distinctions are debatable. Some bioethicists, for example, may consider withdrawing interventions a form of active euthanasia, not passive. Others wonder about the kind and degree of consent necessary to be considered consent, or tantamount to consent. Such skepticism challenges the heuristic value of these categories themselves. Yet the very existence of such conceptual categories disaggregates the range of interventions that could be applied to a dying person. This disaggregation enables bioethicists to analyze each kind and, ultimately, judge the relative morality of each. That is, these distinctions focus the conversation and in normative arguments, focus is vital.

Too narrow a focus, however, can undermine the task of normative argumentation itself. What happens when bioethicists debating a particular form of dying intervention become so fixated upon their point of view that they lose sight of the larger context? The consequences can be grim for patients and families confronting difficult decisions. Such focus can also be detrimental to the textual tradition upon

[7] Byron Sherwin, "A View of Euthanasia." In *Contemporary Jewish Ethics and Morality*, edited by Elliot N. Dorff and Louis E. Newman (New York: Oxford University Press, 1995), 363.

which Jewish bioethicists rely when constructing their normative claims in the first place.

This danger of hermeneutics requires a demonstration. Consider: Central to nearly every piece of Jewish ethical deliberation on euthanasia is the Talmudic story of Chananya ben Teradyon, a second-century Palestinian rabbi burned alive by the Romans for teaching Torah in public, itself a capital crime. This story is critical to Jewish end-of-life deliberations, for how bioethicists read this story will reflect and reinforce their position on euthanasia generally or a particular kind of euthanasia.

According to the story, the Romans attached Chananya ben Teradyon to a pyre, affixed tufts of wet wool to his chest cavity to keep him cooler longer, wrapped the Torah scroll around him, and lit him aflame. Chananya then had three conversations. The first was with his daughter who complained seeing him in this condition. He reassured her that he was in a reasonable place, theologically speaking, and she should take comfort in this conviction. The second was with his students who asked what he saw up there in the flames. He said he saw the parchment of the Torah burn but the letters of revelation fly into the sky. Then they encouraged him to open his mouth to asphyxiate faster. He retorted that one should not injure oneself.

It is the final conversation while Chananya's flesh melts away that best illustrates the dangers of hermeneutics, that is, how reading a text—or a body—is potentially lethal. Before citing that exchange in its entirety, it should be recalled that there are two significant cognitive motivations when looking at the world, including texts. According to social psychologist Ziva Kunda, directionally oriented cognition motivates looking at the world in such ways that are "most likely to yield the desired conclusion."[8] The other kind of cognitive motivation seeks accuracy and leads to the use of cognitive processes and strategies "that are considered the most appropriate" for the task at hand. In short, accuracy seekers endeavor to see the breadth and depth of what is. Goal seekers, by contrast, consider only those data and shreds of evidence that comport with and reinforce predetermined goals. They see the world—and read a text—as they want to see it. All this holds true as well for Jewish bioethicists who read classic religious sources in pursuit of constructing norms: some read texts as they actually are while others read what they want to see in those texts.

Return now to the third conversation between the dying Chananya and his final interlocutor: it is with no other than his executioner.

> The executioner said to him, "Teacher, if I increase the flames and remove the tufts of wool from your heart, will you bring me to the World to Come?" He said to him, "Yes." "Swear to me." He swore to him. Thereupon he immediately increased the flames and removed the tufts of wool from his heart. His soul quickly departed. [The executioner] even jumped and fell into the fire.[9]

---

[8] Ziva Kunda, "The Case for Motivated Reasoning," *Psychological Bulletin*, 108/3(1990):480–498.

[9] BT *Avodah Zarah* 18a (Vilna). Translation by Jonathan K. Crane, *Narratives and Jewish Bioethics* (New York: Palgrave Macmillan, 2013), 102.

This short exchange between a dying person and his executioner is perhaps unsurprisingly read differently by Jewish bioethicists. As I have demonstrated in work elsewhere, Jewish bioethicists most frequently read this text with predetermined goals in mind.[10] That is, they see in this story—and in this final conversation specifically—what they want to see in it. Some bioethicists see in this exchange permission for passive euthanasia, the evidence for which is the plan to remove the tufts of wool. This kind of reading would endorse the contemporary practice of withdrawing interventions from the dying. Others argue that this little chat supports voluntary euthanasia because Chananya agrees to the plan, which would provide Judaic support for the Death with Dignity movement now making its rounds in North America and Europe. Some disagree with this position. For them, this text validates involuntary euthanasia because, in their view, Chananya's supposed consent to the plan that the executioner—not he—devised in the first place was more window-dressing than substantive. And still others enthusiastically point to the fact that the flames were increased as proof to authorize active euthanasia.

In these and so many other ways Jewish motivated bioethicists narrowly read this story—and this particular brief exchange between an enflamed rabbi and his executioner in a public, ghastly scene eighteen hundred years ago—as grounds for contemporary norms regarding intervening in someone's dying process. In their deft hands, they make this brief encounter say what they want it to say. Because of this, they are able to claim that the Judaic textual tradition endorses their position on this or that kind of euthanasia, just as they had anticipated.

Such reading strategies put the proverbial cart before the horse. More, they put that cart and horse on a particular path that leads precisely where the bioethicists want it to end. In doing so, however, bioethicists cut and contort this and other classic resources that are otherwise rich, nuanced, multi-vocal, ambiguous and even ambivalent. For example, a more holistic reading strategy for this particular story would embrace the fact that the central individual, Chananya, changes his attitude while he experiences his own dying. To his daughter he expresses relative calm; he neither requests nor prohibits interventions. To his students he sternly articulates a position that nothing whatsoever should be done to hasten his demise, or more precisely, that he himself will do nothing to bring his death about faster. And yet to his executioner, whose job, mind you, is to kill him, Chananya agrees to a plan to quicken his own dying process, a plan that itself is neither his nor simple. A more robust reading strategy would at least recognize these kinds of experiences and attitudes during the dying process. Such a reading strategy would be akin to Kunda's accuracy-seeking cognition: it reads what is there in the text, and appreciates the full range of its ambivalences and ambiguities. This, of course, complicates the desire to extract a norm from this source. But perhaps that is precisely what is needed especially when confronting the emotionally dynamic and ethically fraught issues of another's dying.

[10] See Crane 2013.

As it is, however, the reading strategies other Jewish bioethicists use are predominantly goal-seeking: they read in the text what they want to see in it. This particular story about Chananya is so rich that it can be read to permit *and* prohibit this or that kind of intervention in dying; it is unsurprising, then, that this story figures so significantly in Jewish discourse on end of life decision-making.

Yet goal-oriented reading, however, is dangerous for several more reasons.

First, it flattens this rich, lively, narrative into an ahistorical norm. By fixating upon a tiny and isolated fraction of the overall story, this kind of reading is akin to slicing a living biological specimen for a slide, and squishing it for convenient analysis, just to extract from this now dead sliver a diagnosis of and for the whole organism. Whatever norm is selected essentially reflects a dead and perhaps deadly interpretation of the story.

Second, bioethicists fixating upon this final exchange between Chananya and his executioner make a disturbing analogy concerning professionals and the dying. On the one hand, central to the executioner's job description is the responsibility to kill his charge. Modern health care providers may rankle at this comparison insofar as many if not most understand their primary duty is to care for if not cure their patients. And this leads to the other disturbing analogy. Chananya was a criminal, at least in the eyes of the Roman regime, and deserved to die, legally speaking. Would this mean today's dying patients deserve to die, and to die at the behest of the government at that? Moreover, Chananya was killed in a dirty public square, exposed to all and sundry, whereas modern dying most frequently occurs in private, if not isolated hygienic settings. Perhaps such disjunctions undermine this story's unquestioned stability as a ground for modern norms.

Third, goal-oriented bioethicists often manipulate the primary source itself to make it say precisely what they want it to say. For example, some reframe or eliminate the executioner's motives (gaining access to the World to Come) to recharacterize him as merciful or as interested only in shortening Chananya's intense suffering. Others have Chananya as the one who comes up with the plan in the first place. Most, however, cut the executioner's plan so that he says either that he will raise the flames or that he will remove the wet tufts of wool from Chananya's chest, depending on whether they endorse active or passive euthanasia, respectively.[11] Like the dying bodies of today's patients that are vulnerable to other's invasive actions, the text itself lays vulnerable to bioethicists' hermeneutical incisions: it, too, is cut, excised, and stitched back together according to others' visions of what it ought to be.

Narrowly focusing on one kind of dying intervention often leads bioethicists into reading strategies that confine thinking and argumentation. It constrains the sources they consider as much as how they read the selected texts. Indeed, they become highly selective of only those details they deem most salient within those few texts. And from these limited resources they then declare wide-sweeping norms

[11] See Crane 2013 for extensive demonstration and analysis of these and other kinds of manipulations.

about what to do vis-à-vis the dying. Such manipulations trouble the trust normative deliberations otherwise merit.

And this leads to the third layer of discursive trouble, the layer of meta-ethics. Even before they start reading texts or writing about them, Jewish bioethicists live and work within particular worldviews. These worldviews or meta-ethical schools value living and dying in distinctive ways, and each cultivates peculiar discursive practices. Some bioethicists, for example, are concerned primarily about outcomes, like prolonging life at all costs, and thus speak predominantly in consequential terms. Others stand behind, upon or under principles, and in their view these principles, like loving one's neighbor, provide solid grounding for pragmatic norms. For them, deontology is the language for thinking about the dying. Some other bioethicists, by contrast, are guided by theology. They assume their theories of covenant or redemption provide necessary and sufficient guidance on what should be done in a peculiar biomedical situation with a particular dying patient. Other bioethicists consider law the most authoritative. Since living overflows with duties, they speak in terms of jurisprudence and precedence. And some bioethicists hold that insofar as lives are unfolding stories, narratives are the rightful source for and instantiation of norms even for the dying. To be sure, even more worldviews speak into the fray.

Each worldview and its discursive practice is, for the most part, internally coherent. And each is relatively healthy insofar as bioethicists sharing the same worldview often read, cite and critique each other's work.

Such internal coherence belies the cracks in the crevice of the conversation as a whole. It is, for example, difficult to discern which worldview's normative position, if any, should prevail when disagreement between them obtains. What are the uncontroversial mechanisms and values that will or should guide deciding which worldview and discursive school should take precedence?

Consider the fractious debate in Jewish circles about the tension between law and lore. On the one hand, some hold that while law offers concrete guidance, it is sterile unless and until it is situated within a larger narrative context. For them, any law, and law itself, makes sense only within a larger, civilizational, story. In their view, narratives—such as stories in the bible or rabbinic midrashim—should be the primary source for norm creation.[12] The other camp contends that law's potency is indisputable, and lore, irrespective of its source, is useful to the degree it reifies the law. For them, moral conundrums are best thought through by legal formalism.[13] Which approach should trump? How can their respective arguments be assessed, by which standards, and who is authorized to make such judgments? While such meta-ethical conundrums are not unique to Jewish bioethical discourse, they nonetheless warrant some attention.

These three discursive layers—linguistic, hermeneutic, and meta-ethical—are not mutually exclusive, of course. They intersect and often amplify their respective difficulties. This can be seen in their disagreements about who is dying, what kind

[12] This is the position of Robert Cover and many other narrativists.

[13] This is the position of Julius Preuss, Emmanuel Jakobovitz and other legal formalists.

of death they are dying, and why their living or dying matters in the first place. Despite the School of Hillel's model of being humble, in this kind of dynamic deliberative context, humility does not guarantee that one's proposed norm will be intelligible to, much less become endorsed by, the contemporary Jewish community that does not share one's primary worldview.

4.4 Advancing Problems

Another arena illustrates the apparent incoherence in Jewish normative deliberation regarding end of life care. This is the arena of advance health care directives. These forms guide individuals and families through many important decisions, like designating a health care proxy, and more difficult ones like intubation, resuscitation, nutrition and hydration. Some forms also contend with the very definitions of death, the permissibility of organ donation, and autopsy. All, however, reflect and reinforce particular notions of who are the rightful authorities to make decisions about end of life care.

Two Orthodox rabbinic associations have published nearly identical advance health care directives: the Rabbinical Council of America, and Agudath Israel. In section two, both forms declare that Jewish law is to govern the health care decisions for the above-signed individual. It reads:

> I am Jewish. It is my desire, and I hereby direct, that all health care decisions made for me (whether made by my agent, a guardian appointed for me, or any other person) be made pursuant to Jewish law and custom as determined in accordance with[14] Orthodox interpretation and tradition. Without limiting in any way the generality of the foregoing, it is my wish that Jewish Law and custom should dictate the course of my health care with respect to such matters as the performance of cardio-pulmonary resuscitation if I suffer cardiac or respiratory arrest; the performance of life-sustaining surgical procedures and the initiation or maintenance of any particular course of life-sustaining medical treatment or other form of life-support maintenance, including the provision of nutrition and hydration; and the criteria by which death shall be determined, including the method by which such criteria shall be medically ascertained or confirmed.[15]

In no uncertain terms, then, signers of these forms desire Judaism to guide not just the decisions about which interventions to apply, maintain, withdraw or withhold. They also want Jewish law and custom to dictate the very criteria of death and the methods by which that criteria are to be assessed. Jewish law and custom not only trump professionals experienced and trained in medicine, Jewish law and custom is to be the lingua franca of the care itself when the patient is Jewish.

Of course, this position puts a particularly heavy burden upon health care proxies. It assumes that proxies will be familiar with Jewish law and custom regarding

[14] In the Agudath Israel document, the word "strict" precedes Orthodox.

[15] "Proxy and Directive with Respect to Health Care Decisions and Post-Mortem Decisions." The Rabbinical Council of America.

care for the dying. It also assumes proxies will be expert enough and dispassionate enough to communicate Jewish law and custom to health care providers in clear and unambiguous terms. In the case that a proxy is unable or unwilling to give such direction to the professionals, the next section instructs that individual to "consult with the following Orthodox Rabbi and I ask my agent to comply with his halakhic decisions.[16]" If that rabbi is unavailable, unable or unwilling to serve in this capacity, an alternate rabbi is also named. The document's instructions for completing this form tell the patient that these named rabbis are the ones "whose guidance you want your agent to follow, should any questions arise to the requirements of halakhah." That is, even identified proxies are not free to make health care decisions on their own for this patient unless and until they have sought the legal counsel of one of these named rabbis. Decisions, at least for Orthodox patients, are to be driven by selected rabbis.

Though patients are free to name any Orthodox rabbi of their own choosing, the form does make a strong suggestion that each rabbi be vetted. The form instructs: "You are encouraged to discuss the matter with the rabbi to ascertain his specialization in end-of-life halachic issues and willingness to serve in such capacity." Such language suggests patients are themselves knowledgeable enough about end of life issues to assess a rabbi's expertise. This presumption is unfounded in many if not most instances, as only fellow experts in these issues would be qualified to make such judgments about a rabbi's command of the material, comprehension of complications, and ability to communicate to health care professionals.

Besides identifying rabbis as the supreme authority, the operating assumption in these Orthodox advance directives is that "all essential emergency and/or life sustaining measures" are to be taken unless and until an authorized proxy declares otherwise. Only after the conclusion of the official form duly signed by witnesses is there an appendix in which expressions of intent are to be clarified vis-à-vis various and specific end of life medical decisions. These include such scenarios as being terminally ill, being in a coma with no hope of recovery, being so brain damaged or diseased that the patient cannot recognize people or speak and no improvement is possible. That these scenarios are an afterthought, literally and figuratively, reinforces the underlying presumption that quantity of life trumps quality.

At perhaps the other extreme is the Reform community's commitment to deferring to individuals for determining what should and should not be done to them when dying. In a 1991 resolution, the Union of American Hebrew Congregations "reaffirm that in accordance with Jewish tradition each individual has the ethical, moral and legal right to make his own or her own health care decisions, and that such right survives incompetency."[17] That resolution encourages congregations to develop educational programs for community members to complete legally acceptable and binding advance care documents. By democratizing this effort, the UAHC

---

[16] This reads, "I ask my agent to follow his guidance" in the Agudath Israel document.

[17] Union for Reform Judaism, http://www.urj.org/what-we-believe/resolutions/health-care-decisions-dying, accessed September 6, 2017.

and its rabbinic association, the Central Conference of American Rabbis, simultaneously abdicated the responsibility of providing a centralized document and deferred to congregation leaders to decide how Judaism would play in their educational fora and even in the advance directive forms they promoted.

For example, the Kalsman Institute on Judaism and Health, housed at the Reform seminary Hebrew Union College-Jewish Institute of Religion, links to what it calls a "Reform Movement Advance Directive Form."[18] That form is a revision of a document prepared by Linda and Ezekiel Emanuel and published in the *Journal of the American Medical Association*.[19] That form opens with the statement, "as part of a person's right to self-determination, every adult may accept or refuse any recommended medical treatment." The authority of the individual is indisputable: all should defer to patient autonomy.

Structurally, the form begins by identifying six scenarios in which the patient is incompetent: permanent coma, near death, weeks to live cycling in and out of consciousness, extreme dementia, temporary inability to make decisions, and something the individual is to describe. For each one, goals and interventions must be selected. Identified goal options include prolong life, attempt to cure and reevaluate often, limit to less invasive and less burdensome interventions, and provide comfort care only. The patient then must complete a matrix of various interventions, marking those which are wanted, those which are not wanted, and those about which the patient is undecided. Only at the end of the document are proxy decision-makers identified; indeed, a patient may assign co-proxies. The proxy is directed to make decisions based on his/her assessment of the patient's wishes as articulated in the completed matrices in the aforementioned six scenarios. Absent those, the proxy is bidden to decide based on his/her best guess as to the patient's wishes. And if there is a conflict between the patient's goals, selected treatments, and final personal statement, the form asks the patient to identify which of these should be given greater weight.

Such deference to the wishes of the individual patient reflects and reinforces modernity's fascination with personal autonomy, and Reform Judaism's too. That said, the form's reference to religion is negligible, its mention of Judaism nil, and rabbis are nowhere to be found. The ultimate authority, then, is the modern patient who may—or may not—be guided by Judaism.

Nestled between these two extremes is the Jewish Medical Directives for Health Care form developed by the Committee on Law and Standards of the Rabbinical Assembly for the Conservative movement.[20] Its instructions begin with the acknowledgment that advances in medicine require making known our preferences to others, and that "we should be guided by our commitment to Judaism, to its law

[18] Kalsman Institute on Judaism and Health, http://kalsman.huc.edu/articles/URJ,%20Jewish%20Family%20Concerns%20Advance%20Directive.pdf, accessed September 6, 2017.

[19] "The Medical Directive: A New Comprehensive Advance Care Document." *JAMA*, 261:3288–3293, June 9, 1989.

[20] The Rabbinical Assembly, http://www.rabbinicalassembly.org/story/jewish-medical-directives-health-care-living-will?tp=178, accessed September 6, 2017.

(halakhah), and to its moral values." Its two forms—a proxy directive and instruction directive—provide legally binding power to the proxy as well as clarity on the kinds of treatments the patient desires in certain scenarios. The identified treatment options "have been constructed in accordance with Jewish law and values, as interpreted within Conservative Judaism."

The instructions then articulate Jewish perspectives on health care. It invokes the notion that humankind is created *b'tselem Elohim*, in God's image, and this bestows each with intrinsic dignity and value in God's eyes. This is then interpreted to mean that "we have the responsibility to care for ourselves and seek medical treatment needed for our recovery," which puts into words what is assumed in the Orthodox form that life must be preserved at nearly all or any cost. On the other hand, though the document declares that Judaism rejects "any form of active euthanasia or assisted suicide", it does acknowledge that each patient must choose "what mode of treatment he or she feels would be both beneficial and tolerable"—which echoes the turn to the individual championed in the Reform movement.

This form then makes clear that it integrates two differing positions about death and dying and interventions therein. One position is by Avram Reisner and the other by Elliot Dorff, both respected Conservative rabbis, legal scholars and bioethicists. While they share many foundational assumptions about health care and decision-making, they disagree about the relative latitude individuals may have when choosing between certain options in limited scenarios. Deciding between these options requires, as the form's instructions say, consulting with "your rabbi to discuss the values and norms of Jewish ethics and halakhah." Guidance may also be sought from physicians, attorneys, and state laws. That is, a person's decisions may reflect extra-Judaic sensibilities.

Understandably, then, conflict may arise between what an agent judges a patient's wishes would be and how someone else interprets the instructive directives. While Reisner and Dorff disagree which should trump in these situations—the agent or the instructive directives, the form includes a section in which the patient must identify the ultimate authority. The Health Care Proxy form concluding the Reform packet has a similar opportunity for the patient to identify the proxy's judgment or the written preferences as the ultimate authority. That is, the individual patient predetermines the rightful authority for when the situation becomes particularly thorny.

The Conservative instruction directive opens with a declaration that the patient is a Jew and that Jewish teachings and values guide and inform how that patient has lived and wants to die. "I share Judaism's respect for my body, [itself] the creation and possession of God, and I consequently wish that all prudent medical treatment be extended to me with the aim of effecting my recovery. Nothing in this directive should be construed as a wish to die…I unequivocally reject any form of active euthanasia or assisted suicide." The prudence mentioned here counters the Orthodox call for "all essential emergency measures" and echoes the Reform insistence for "medically reasonable" treatments. On the other hand, the unambiguous rejection of active euthanasia and assisted suicide places substantial distance between the Conservative and the Reform movement's comparatively moderate attitude toward them. In brief, the Conservative advance directive hews close to Jewish conservation of life while at the same time retaining a sense of personal empowerment for each individual to make personal decisions.

So who, ultimately, is the rightful authority to make decisions regarding end of life care? It depends whom you ask. For Orthodox Jews, it will inevitably be a designated rabbi. For Reform Jews, the patient is that individual. For Conservative Jews, it is the patient whose choices have been circumscribed by Jewish tradition and values, choices that have been identified by rabbis steeped in bioethical discourse.

4.5 Boundary Problems

Without digressing too far, a further complicating feature of this question about Jewish norm-genesis regards the edges of those norms. To what extent do Jewish norms extend to non-Jews? And in which ways might Jewish norms (and Jews themselves) be impacted by non-Jewish influences?

Consider, for but one example, that Emanuel Jakobovits, then the Orthodox Chief Rabbi of Great Britain, felt compelled by the emergence of Catholic bioethical discourse to develop a Judaic resource on the topic. His *Jewish Medical Ethics: A Comparative and Historical Study of the Jewish Religious Attitude to Medicine and its Practice*, published in 1959, certainly addressed many pressing biomedically and morally complex issues, and for this it marked a watershed in contemporary Jewish ethics for opening up the field of bioethics for Jewish deliberation. But perhaps even more important about this book is that Jakobovits relies on more than Jewish law, as would be expected of an Orthodox scholar; he turns to and integrates a range of Christian, Greek and contemporary scientific resources to make his arguments. That is, even Orthodox Jewish norm deliberation need not be bound to Jewish sources alone.

The classic rabbinic notion of *dina d'malkhuta dina* (the law of the land is the law) offers a more conceptual appreciation of the boundary between Jews and non-Jews. Built upon Jeremiah's call to "seek the peace of the city" into which Jews had been exiled (Jeremiah 29:7), *dina d'malkuta dina* stipulates that Jews are to abide by a regime's rules as long as those rules are explicit and universal. In most matters gentile law would supersede Jewish law; exceptions held in regard to religious issues, when Jewish law would trump. In broad terms and in many areas of life, this concept demonstrated Jewish acquiescence to non-Jewish norms.

If that is on the side of ingress, of gentile influence shaping Jewish norms, what about egress: in which ways might Jewish norms apply to those beyond Judaism? Here, as elsewhere, Jews disagree. Some hold that Jewish norms should be applied even to gentiles, whereas others insist that Jewish norms are relevant only to fellow Jews. Should, for example, Jewish doctors enact Jewish sensibilities for non-coreligionists? What if a patient asks for a procedure that Judaism abhors: should a doctor comply with the request? Or because of this moral distress, should a doctor refuse and/or recuse? Answers, of course, vary.

4.6 Concluding Reflections

Jewish bioethical deliberation on death and dying suffers even in its dynamism. Some might say that it is an incoherent conversation insofar as there is fundamental disagreement about what some of its terms are, such as the central concept of the *goses*. This is further complicated by the fact that there is no standard, uncontroversial way of reading classic sources, and this hermeneutical diversity skews old stories and possibly skewers modern bodies. And when worldviews collide, no clear direction is available to discern which should trump.

Similarly, there is no monolithic approach to guiding Jews to make decisions regarding end of life care. Advance directive forms promulgated by the three major streams of modern Jewry dramatically disagree about who or what should be the rightful ultimate authority. One stream thinks individuals unencumbered by ancient traditions or laws are the rightful authority. Another could hardly disagree more: tradition overrides everything, and that tradition is best instantiated by a rabbi, hopefully one that is expert in such matters. And the third stream meanders between these two, envisioning a Jewishly educated individual as the rightful authority.

Given such deep discursive and pragmatic difficulties and oddities, what can be said about creating norms in contemporary Jewish bioethics? Recall the multi-year debate between Shammai and Hillel. Some might say that the only norm to consider is the one they advocate. Others might say that there are so many options to consider and each is more or less equally valid.

Those like Shammai, think their norm is the right one for modern Jews. Primarily for ego or doctrinal reasons they dismiss both the existence and validity of countervailing definitions and positions. Though perhaps repugnant for such dishonesty, providing decisive and exclusive answers to morally probing questions provides clear guidance, guidance dearly needed and desired especially when making difficult decisions about another's dying. Another strategy would be more like kind Hillel, who taught the myriad range of possibilities before advocating a particular one. While honest that there are competing valid options, this strategy would be exhausting and confusing for audiences insofar as they are already in moral distress about their circumstances and now they have to endure this litany of potential ways to proceed. It remains unclear whether in Jewish bioethics it is better to be direct and dishonest or kind and confusing.

In sum, these two arenas regarding speaking about and authorizing dying thus illustrate some of the concerns regarding normativity in contemporary Jewish bioethics. Discursive and pragmatic discontinuities do not mean, of course, that these efforts are vapid and worthless. On the contrary. They point to the challenges Jews face when constructing norms for their communities, insofar as those communities are embedded simultaneously in a modern sociopolitical context, in an ancient and evolving tradition, and in an ever innovating medical environment.

Suggested Readings

On Jewish Ethics:

Crane, Jonathan K. 2015. Jewish Ethics. *Jewish Studies Bibliography Online*. Oxford University Press. http://www.oxfordbibliographies.com/view/document/obo-9780199840731/obo-9780199840731-0087.xml. Accessed 6 Sept 2017.

Offers the most recent annotated bibliography of the field.

Dorff, Elliot N., and Jonathan K. Crane, eds. 2013. *The Oxford Handbook of Jewish Ethics and Morality*. New York: Oxford University Press.

A state of the art survey of the field, complete with fresh essays by leading scholars around the world on a range of historical, methodological and topical issues.

Dorff, Elliot N., and Louis E. Newman, eds. 1995. *Contemporary Jewish Ethics and Morality*. New York: Oxford University Press.

Collection of classic essays on meta-ethical, methodological and topical concerns.

Newman, Louis E. 1998. *Past Imperatives: Studies in the History and Theory of Jewish Ethics*. Albany: State University of New York Press.

A description of the field, critically analyzing the relationships between ethics, law, and theology.

On Jewish Bioethics:

Bleich, J. David. 1998. *Bioethical Dilemmas: A Jewish Perspective*. Hoboken: Ktav.

A compilation of previously published essays on a range of classic bioethical issues, all treated from an Orthodox perspective.

Dorff, Elliot N. 1998. *Matters of Life and Death: A Jewish Approach to Modern Medical Ethics*. Philadelphia: Jewish Publication Society.

A cohesive book surveying a range of biomedical issues; the appendix outlines Dorff's understanding of the relation between religion and morality.

Jakobovits, Immanuel. 1959. *Jewish Medical Ethics: A Comparative and Historical Study of the Jewish Religious Attitude to Medicine and its Practice*. New York: Bloch.

A classic, perhaps the first major work in the field, drawing together classic Jewish and other sources on a range of biomedical topics.

Zohar, Noam, ed. 2006. *Quality of Life in Jewish Bioethics*. Lanham: Lexington.

Offers a range of Jewish positions on the concept and pragmatic implications of "quality of life" in medicine, especially at end of life.

On Jewish views of Euthanasia:

Crane, Jonathan K. 2013. *Narratives and Jewish Bioethics*. New York: Palgrave Macmillan.

This analysis of a classic Talmudic source critical to contemporary Jewish euthanasia scholarship reveals some shocking reading strategies.

Hurwitz, Peter J., Jacques Picard, and Avraham Steinberg, eds. 2006. *Jewish Ethics and the Care of End-of-Life Patients: A Collection of Rabbinical, Bioethical, Philosophical, and Juristic Opinions*. Jersey City: Ktav.

A collection of essays drawn from such fields as law, history, philosophy and clinical medicine.

Mackler, Aaron L. 2000. *Life and Death Responsibilities in Jewish Biomedical Ethics*. New York: Jewish Theological Seminary of America.

Essays by the Conservative movement's Committee on Jewish Law and Standards, especially on end of life care, physician assisted suicide, and euthanasia.

Chapter 5
Using the Imagination in Normative Moral Reasoning Around the Principle of Double Effect to Foster Doctrinal Development in Catholic Bioethics

Gerard Magill

Abstract The essay explores how the imagination contributes to moral normativity. The purpose is not to argue that the imagination provides a particular mode of reasoning. It does not. Rather, the imagination can present riveting images based on practical experience that facilitate reasons to coalesce around a conclusion. This presents what can be seen as a common sense view of moral normativity. The experience of marriage or belief in God can be illustrative of reaching certitude even when demonstrative proof is not feasible. In this process, the imagination must always be accountable to reasoning. Also, the dynamic nature of experiential images can inspire action, contrasting with abstract modes of moral normativity that may appear detached from accompanying action. The analysis suggests that the abstract process of formal reasoning adopted in the classical principle of double effect can be enhanced by a role for the imagination. The role facilitates personal and practical reasoning to coalesce around insight and moral conclusions that can be explained further using the abstract process of double effect reasoning. This approach can foster doctrinal development in Catholic bioethics. The religious epistemology and theological method of John Henry Newman (1801–1890) are discussed to provide a theoretical explanation of this interaction between the imagination and reasoning as a distinctive contribution to moral normativity.

Keywords Catholic Bioethics · Church Tradition · Conscience · Imagination · Double Effect · Doctrinal Development · Moral Dilemmas · Normative Moral Reasoning · Religious Epistemology

---

G. Magill (✉)
Center for Healthcare Ethics, Duquesne University, Pittsburgh, PA, USA
e-mail: magillg@duq.edu

J. Gielen (ed.), *Dealing with Bioethical Issues in a Globalized World*,
Advancing Global Bioethics 14, https://doi.org/10.1007/978-3-030-30432-4_5

5.1 Introduction

Normative discourse in Catholic bioethics engages a global perspective by establishing universal principles that are applied in different ways depending on historical and cultural circumstances. Examples of these principles include upholding the dignity of each person and respecting unborn life. When these broad principles are applied concretely they can clash in situations that raise complex conflicts in practical life. The Catholic tradition developed the principle of double effect as a mode of normative reasoning to resolve these dilemmas. However, exploring that principle can help to shed light on another approach called imaginative discernment. The analysis begins by considering how the Catholic tradition addresses situations or dilemmas where there is a clash of values.

5.2 Situations of Moral Dilemmas

In the Catholic tradition, situations of moral dilemmas where there is a clash of values can be addressed by the principle of double effect. In the analysis below, the principle is applied to the dilemma raised by the Phoenix case in a Catholic hospital in the U.S.

5.2.1 Principle of Double Effect

It can help to begin the discussion by recalling the debate between the Natural Law as applied in a universal manner by the Magisterium of the Catholic Church and the personal or historicist interpretation of it in what has become known as Proportionalism. The Magisterium refers to the formal teaching authority of the Catholic Church referring to the universal authority of the Pope, such as in a Papal Encyclical, or of the Pope working with the Catholic Bishops, such as in a Vatican Council or Synod of Bishops. On the one hand, the stance of the Magisterium was explained in the encyclical *Veritatis Splendor* (the *Splendor of Truth*), which was promulgated in 1993 and highlighted "the universality and permanent validity of its [i.e. the Catholic Church's doctrine regarding the natural law] precepts" based upon "reason … proper to human nature."[1] The encyclical contrasted this to the historicist endeavors of Proportionalism: "by weighing the various values and goods being sought, [proportionalism] focuses … on the proportion acknowledged between good and bad effects of that choice," with the result that "it is never possible to formulate an absolute prohibition of particular kinds of behavior."[2] Despite the

---

[1] Pope John Paul II (1993), number 4.

[2] Pope John Paul II (1993), number 75.

Encyclical, Proportionalism as a normative method in Catholic morality continues, prioritizing historicism over absolutism. Rather than going into the details of this ongoing debate,[3] it is sufficient for the purposes of this paper to note that the tension between absolutism and historicism in moral normativity was played out prominently via discussions of the principle of double effect (e.g., Mangan, Knauer, Janssens, Fuchs, Keenan, and others).[4] Rather than re-litigate the minutiae between the absolutist and historicist views of this principle, the discussion turns to a pivotal case in U.S. bioethics. The aim is to explain how common sense discernment can emerge in a reliable manner by using imaginative discernment.

5.2.2 *The Phoenix Case*

The case dealt with the death of an 11-week old fetus in 2009. The mother had a prior condition of pulmonary hypertension and an unanticipated pregnancy was confirmed at seven and a half weeks. A month later, the patient was brought to the emergency room at a Catholic facility, St. Joseph's Hospital in Phoenix, Arizona. The diagnosis of severe pulmonary arterial hypertension led to right-sided heart failure and cardiogenic shock.[5] The placenta had exacerbated the patient's prior condition into a life-threatening circumstance. The Ethics Committee approved an emergency dilation and curettage procedure that ended the pregnancy with the mother's consent. Subsequently (approximately, a year later), the bishop of the Phoenix diocese, based on this case and in the context of other practices at the hospital, revoked the name "Catholic" according to the Catholic Church's *Code of Canon Law*, canon 216.[6] In turn, the health system, Catholic Health West, forfeited its formal identity as a Catholic organization. Sadly, the death of the embryo was unavoidable; but the loss of the hospital organization to Catholic health care was avoidable. Each side, the Bishop and the hospital, appealed to the U.S. Bishops' *Ethical and Religious Directives for Catholic Health Care Services*.

The bishop focused on this directive:

> Abortion (that is the directly intended termination of pregnancy before viability or the directly intended destruction of a viable fetus) is never permitted. Every procedure whose sole immediate effect is the termination of pregnancy before viability is an abortion.[7]

In contrast, the hospital focused on this directive:

> Operations, treatments, and medications that have as their direct purpose the cure of a proportionately serious pathological condition of a pregnant woman are permitted when they

---

[3] Kelly et al. (2013).

[4] Mangan (1949); Knauer (1967); Janssens (1972); Keenan (1993); Kaczor (1998).

[5] Lysaught (2011), 538; Touhey (2011), 8.

[6] Code of canon Law (1983).

[7] U.S. Bishops (2009), number 45.

> cannot be safely postponed until after the unborn child is viable, even if they will result in the death of the unborn child.[8]

This case can be engaged using the principle of double effect, as discussed by the author elsewhere.[9] Agreement on this case can be reached using both the absolutist method, that applies the principle with its traditional four conditions, and the historicist method of Proportionalism that coalesces those traditional conditions into a more streamlined manner. Both can justify the evacuation of the womb to remove the offending organ, the placenta, even though the death of the fetus is foreseen but not intended–recalling the settled case where Church teaching justified a hysterectomy to treat a pregnant woman's cancer. In this situation or case, Pius XII permitted double effect:

> The reason is that, if, for example, the safety of the future mother, independently of her state of pregnancy, might call for an urgent surgical operation, or any other therapeutic application, which would have as an accessory consequence, in no way desired or intended, but inevitable, the death of the foetus, such an act could not be called a direct attempt on the innocent life. In these conditions the operation can be lawful, … provided … that it is not possible to postpone it till the birth of the child, or to have recourse to any other efficacious remedy.[10]

This teaching of Pius XII was re-affirmed in 2010 by the Committee of Doctrine of the U.S. Bishops. The committee explained that "the surgery does not directly target the life of the unborn child."[11]

The mistaken approach to moral normativity in the stance of the Bishop of Phoenix may be caused either through unawareness of how subtle the Catholic moral tradition can be, or, because of a naïve understanding of the pro-life stance of the Catholic Magisterium. The former may require extensive education on Catholic morality, but the latter is a matter of basic understanding about how the Bishops have negotiated arrangements regarding abortion. For example, the U.S. Bishops used a principle akin to double effect, called the principle of material cooperation, to justify their acceptance of the famous Hyde Amendment in the U.S. Government that includes funding for abortion in some cases. The U.S. Bishops accepted the continuation of the 1976 Hyde Amendment (via an annual appropriation bill in the U.S. Congress) during the 2010 debate over the Affordable Care Act of President Obama. Their stance was based on the explicit teaching of the Vatican. In the 1995 Encyclical titled *The Gospel of Life*, Pope John Paul II applied the principle of cooperation to Catholic politicians faced with abortion legislation.

> A particular problem of conscience can arise in cases where a legislative vote would be decisive for the passage of a more restrictive law, aimed at limiting the number of authorized abortions, in place of a more permissive law already passed or ready to be voted on. Such cases are not infrequent…. In a case like the one just mentioned, when it is not

[8] U.S. Bishops (2009), number 47.

[9] Magill (2011).

[10] Pius XII (1951), 859; Liebard (1978), 127.

[11] U.S. Bishops (2010).

> possible to overturn or completely abrogate a pro-abortion law, an elected official, whose absolute personal opposition to procured abortion was well known, could licitly support proposals aimed at limiting the harm done by such a law and at lessening its negative consequences at the level of general opinion and public morality. This does not in fact represent an illicit cooperation with an unjust law, but rather a legitimate and proper attempt to limit its evil aspects. [12]

The purpose of this discussion is not to engage the debate on the relation between the principle of double effect and the principle of cooperation in Catholic moral normativity, though that is a fascinating discourse to undertake. For example, some scholars claim that cooperation is an application of double effect; others argue that the two principles are distinct because double effect involves one agent with two acts, whereas cooperation involves two agents with their separate actions.[13] While not much can be done about the stance of the Bishop in the Phoenix case, an important lesson can be learned from the overlapping agreement between the absolutist and historicist approaches that justified the evacuation of the womb.

The point to be learned is this: two opposing frameworks in moral normativity reached agreement on the Phoenix case. The obvious question is what caused this agreement, especially on such a pivotal case. My argument is that a related method of moral normativity makes an imaginative discernment that the opposing frameworks ratify. This introduces the role of the imagination into discourse about moral normativity.

5.3 Imagination in Moral Discourse

A role for the imagination in moral discourse has been discussed by many scholars in philosophy and theology.[14] The analysis here focuses upon the contribution of the imagination to moral normativity and to doctrinal development in the Catholic tradition.

5.3.1 Imagination & Moral Normativity

The imagination plays a significant and often determinative role in moral discernment. Arguably, it is the imagination that enables the opposing absolutist and historicist approaches in the Phoenix case to reach a common sense discernment. There is a long historical tradition about common sense theory in normative morality, from the secular writings of Thomas Reid (1710–1796) in the Scottish Enlightenment, to

[12] Pope John Paul II (1995), number 73.

[13] Kelly et al. (2013), chapter 24.

[14] Abelson (1977); Gardner (1982); Johnson (1987); Johnson (1993); Kaufman (1981); Lovibond (1983); Mackey (1986); McIntyre (1987); Ricoeur (1995); Tracy (1981); Warnock (1976).

the religious works of John Henry Newman (1801–1890) in the English Victorian age. Newman integrated the stance of Reid with his religious epistemology on the imagination, and this study engages Newman's insights.

To understand common sense discernment, it can help to refer to a broader tapestry, shifting from moral normativity to the justification of religious belief more generally. The core argument is that we rely on the imagination to justify belief in God ("imago dei" being a function of the imagination) across many religions and cultures. Typically, this belief is held with certitude despite the manifest absence of demonstrative proof. Yet this belief is construed as being sufficiently reliable to guide individuals for their lifetime and Churches for millennia. If this belief and its reliance on the imagination can be adequate for these commitments, this approach can also be helpful for making practical ethical decisions especially when faced with heart-wrenching dilemmas, such as in the Phoenix case.

The point is not that the imagination is a form of reasoning: it is not. But the imagination can provide riveting images based on experience that enable reason to grasp practical and complex reality in concrete moral discernment. How might the imagination function in the Phoenix case? The tragic image of losing two lives needlessly appears to drive common sense to perceive the urgency of saving the life that can be saved. Subsequently, different normative frameworks (absolutist or historicist) are used to ratify the discernment.

This common sense discernment arises when a congruence of significant and relevant factors (Newman called these probabilities) coalesce to a point of plausibility or meaning. Just as with belief in God, so with the decision to marry one's spouse: we cannot prove the inference, but a coalescence (Newman referred to this as convergence) of significant and relevant factors occurs to justify moral discernment. Admittedly, this is a very personalist and historicist approach to objectivity in moral normativity that emphasizes there is no subject-free objectivity in moral normativity. If this view of imaginative discernment is sufficient for life-changing moral decisions, such as belief in God or marriage, it should be sufficiently reliable for moral normativity in daily life.

In sum, the imagination provides us with riveting images based on experience that enable us to rationally coalesce significant and relevant factors to reach common sense discernments that are sufficiently justifiable to both act upon and shape moral normativity. These common sense imaginative discernments must be ratified rationally, even by opposing frameworks such as the absolutist and historicist approaches to double effect. Hence, imaginative discernment is always accountable to reasoning. It is also worth noting that this explanation of the imagination not only contributes to discourse on moral normativity but also contributes to a nuanced understanding of doctrinal development, such as in the Catholic moral tradition.

5.3.2 Imagination & Doctrinal Development

On February 18, 2016, during his return trip to Rome from visiting Mexico, Pope Francis indicated to journalists that contraception may be justified for protection against the Zika virus (a mosquito-borne disease). The virus is connected with microcephaly, leading to severe brain damage of the newborn. This can create a public health emergency. The Pope reiterated Church teaching that all abortion is wrong but that avoiding pregnancy is not an absolute evil. He specifically referred to Pope Paul VI permitting nuns in the Belgian Congo to use contraception when in danger of rape. On February 22, 2016, in an interview with Vatican Radio, the Vatican spokesman Fr. Federico Lombardi confirmed this stance of Pope Francis explaining the possible justification of contraception in emergency situations after serious discernment of conscience. This news will be surprising for many and will lead to the question about what sort of "contraception discernment" might be appropriate?

If Pope Francis's stance is to be consistent with the traditional teaching of *Humanae Vitae*, the 1968 encyclical of Pope Paul VI that prohibited using artificial contraception, a nuanced explanation of contraception discernment is needed. In his remarks there appears to be an application of the situation of the Congo nuns (intending contraception in a circumstance that did not involve freely chosen intimacy between spouses) to a circumstance that does involve freely chosen intimacy between spouses. The problem is that forbidding the latter seems to have been the purpose of *Humanae Vitae* (the former not being within its purview) – referring to "the unitive significance and the procreative significance which are both inherent to the marriage act" the Church forbids "any action which either before, at the moment of, or after sexual intercourse, is specifically intended to prevent procreation–whether as an end or as a means."[15]

The contraception discernment that is necessary, as mentioned by the Vatican spokesman, can be guided by double effect reasoning that appears in the paragraph in *Humanae Vitae* immediately following the prohibition of artificial contraception: "On the other hand, the Church does not consider at all illicit the use of those therapeutic means necessary to cure bodily diseases, even if a foreseeable impediment to procreation should result there from–provided such impediment is not directly intended for any motive whatsoever."[16] This justification of contraception as a therapeutic means for bodily disease highlights the remark of Pope Francis that avoiding pregnancy through contraception is not always evil. This mode of moral reasoning applies the traditional principle of double effect and it provides the guide rails for the sort of contraception discernment that Pope Francis urges. The Pope's remarks, confirmed by Fr. Lombardi as the Vatican's official spokesman, caused a robust conservative reaction in Catholic circles in the US. Two interesting points need to be noted to advance this discourse from the perspective of moral normativity.

[15] Pope Paul VI (1968), number 12 & 14.

[16] Pope Paul VI (1968), number 15.

On the one hand, although the Pope did not refer to using double effect reasoning to justify his stance, his call for discernment requires the sort of detailed analysis that the principle provides. It is worth rehearsing briefly how even the absolutist approach to the principle of double effect might function here to justify contraception as a protection against the Zika virus. The standard four conditions of the principle can be applied. First, the action itself must not be morally wrong–as Pope Francis emphasized, preventing pregnancy is not always evil. Second, the negative effect (in this case using artificial contraception to prevent pregnancy) must not cause the good effect (in this case, protection from the Zika virus)–at worst in this scenario these occur simultaneously and that is permissible. Third, the individual must intend the good effect (protection from the Zika virus) and not the negative effect (prevention of pregnancy). Here the crucial difference between intention and foresight is indispensable–the knowledge that using contraception will inevitably prevent pregnancy does not mean that preventing pregnancy is the intent. Fourth, there must be an appropriate balance or proportion between the good effect and the negative effect–that appears to be the case when contraception provides an effective protection against the Zika virus. Despite conservative contestation, the Pope's stance can be construed as a function of imaginative discernment that can be ratified by both absolutist and historicist approaches to double effect.

On the other hand, it is important to grasp what seems to have led the Pope to make such a significant statement that enables couples to prevent pregnancy using artificial contraception (the very point prohibited in *Humanae Vitae*)–imaginative discernment. The leitmotif of mercy that characterizes the papacy of Francis seeks to bridge abstract norms or doctrine with practical reality and capability. By doing so, the Pope appears highly attuned to the cruel image of microcephaly that involves severe brain damage of the newborn, combined with the heart-wrenching reality of families having to care for them throughout their lives. That haunting image drives us to avoid the viral infection that leads to the disease. It appears as common sense to use artificial contraception insofar as it provides effective protection. The traumatic image of microcephaly enables populations to reach the common sense discernment that we must take effective measures, especially when they are cheap and available, to prevent the dreadful disease and its awful impact upon newborn. Subsequent to this imaginative discernment both absolutist and historicist approaches to double effect can ratify this common sense insight.

The imagination of Pope Francis enabled him to enact his commitment to mercy by privileging the practical discernment of parents over the abstract constructs of prior doctrine. His common sense discernment based on the imagination (via the stark image of microcephaly) can be ratified using abstract frameworks like double effect. By presenting his common sense discernment, the Pope shed light on how the teaching of *Humanae Vitae* can be applied to devastating scenarios that could not have been anticipated previously. His stance, explained here using imaginative discernment as a form of moral normativity, also sheds light on how the teaching of *Humanae Vitae* needs to adapt, change, and develop to accommodate emerging catastrophes like the Zika virus.

Another example of the imagination by a different Pope led to a very surprising development of doctrine in the Catholic Church. It may appear that doctrinal development occurs slowly over centuries. But that would be erroneous. Doctrine also can develop suddenly, albeit reflecting years of prior experience and deliberation. The example here is that of the teaching of Pope John Paul II on capital punishment when he effectively ended any toleration by the Catholic Church for executing criminals. For centuries the Church had accepted the legitimacy of executing criminals. Over many decades in the twentieth century there had been considerable debate about the morality of this practice. That ongoing discussion had such an impact on Pope John Paul II that suddenly, to the surprise of most in the Church especially so-called socially conservative bishops, he changed Church teaching. In his 1995 encyclical, the *Gospel of Life*, he changed Catholic moral doctrine: he prohibited the death penalty for criminals. He conceded only one exception, when a country could not protect its citizens from a murderer: "... punishment ... ought not go to the extreme of executing the offender except in cases of absolute necessity ... when it would not be possible otherwise to defend society .... Such cases are very rare, if not practically non-existent" (referring also to the *Catechism of the Catholic Church*).[17] To explain his stance, the Pope referred to a well-known image: "to kill a human being, in whom the image of God is present, is a particularly serious sin."[18] Here again we see the role of imaginative discernment that coalesces reasons toward a practical moral conclusion.

Another example of doctrinal development might be worth considering as potentially occurring in the future via imaginative discernment. This example deals with Church doctrine on ordination, but it has implications in morality from the perspective of gender equality. In the Roman Catholic tradition, women have not been eligible for ordination, either as deacons, priests, or bishops. Specifically, Pope John Paul II published an Apostolic Letter on "Priestly Ordination" to "declare that the Church has no authority whatsoever to confer priestly ordination on women and that this judgment is to be held definitively by all the Church's faithful."[19] Following this teaching, on November 1, 2016, on a flight returning from a Papal visit to Sweden, Pope Francis intimated that the ban on priestly ordination for woman could continue forever. Nonetheless, despite the stark nature of this prohibition, Pope Francis suggested an openness to the ordination of women to the diaconate. Previously, in summer 2016, he had instituted a 12-member "Study Commission on the Women's Diaconate" (with an equal number of women and men on the Commission). The purpose of the Commission is to study the possibility of women being ordained as deacons in the Catholic Church. Perhaps the image of women serving in ordained ministry in the Catholic Church struck the imagination of Pope Francis. This image may have driven him to seek reasons to justify his common sense intuition, especially at a time in contemporary society when cultural bias against women is widely

[17] Pope John Paul II (1995), number 56; Catechism (2000), number 2267.

[18] Pope John Paul II (1995), 55.

[19] Pope John Paull II (1994), number 4.

discussed. If Pope Francis finds sufficient rationale to accept this imaginative discernment regarding a woman diaconate, that would involve a considerable development of doctrine. Furthermore, it could very well be that a future Pope or Council would be able to advance the matter further, such as appointing women Cardinals who would elect future Popes.

It is certainly fascinating to see different Popes take stances that align with Newman's sense of imaginative discernment: this sheds light on how imaginative discernment contributes both to moral normativity and to doctrinal development in religious traditions. The sections below examine Newman's thought to shed light on this role of the imagination in moral normativity.

5.4 Imagination in Newman's Religious Epistemology

To provide a deeper understanding of the importance of imaginative discernment it can be instructive to explore in some detail the religious epistemology of John Henry Newman, upon which the above explanation is based. At nearly 70 years of age he published his renowned work on the philosophy of religion, *An Essay in Aid of a Grammar of Assent*.[20] This was his final explanation of an already sophisticated stance that he had presented in his Oxford University Sermons (1826–1843) as a young Anglican Vicar.[21] At the core of his argument is an explanation of how reason and imagination must work integrally together. He presented this integration, described as "real ratiocination and present imagination,"[22] as normative for both religious belief and religious morality. This integration of reason and imagination can be construed as his hermeneutics of the imagination. To understand this, it is necessary to explain the connection between inference and the assent of certitude in his writings.

5.4.1 *Inference and Certitude*

Newman was adamant that the imagination must always answers to reason: "the imagination … should always be under the control of reason."[23] Upon this foundational insight he built his religious epistemology on these inter-connected components: a subjective form of reasoning that is sufficient to justify religious belief or morality (informal inference); and the discernment of belief or morality as objectively true in the assent of certitude.

---

[20] Newman (1870a).

[21] Newman (1843).

[22] Newman (1870a), 316.

[23] Newman (1870a), 121.

His theory of knowledge had these stages; there are intuitive or inductive apprehensions from particular experiences that elicit simple assent; abstractions are formed from these experiences to elicit what he called notional assent; and these abstractions are applied to concrete conclusions to elicit what he called real (practical) assent. He illustrated this process by reference to morality: first, apprehensions arise from concrete experiences like an act of cruelty or justice; second, from these experiences we generalize an abstract proposition like the rule of right and wrong or a code of moral duties; third, those abstractions are applied to concrete situations indicating that a particular action is right or wrong.[24] Aligned to this process, he made a crucial distinction between conditional and unconditional truth claims. Reasoning involves conditional inferences in the sense that the conclusion is dependent upon the inferences. However, when the conclusion is justified it can be held independently from the inferences as an unconditional assent (called certitude).[25] The outcome of the process was to justify certitude in religious belief and morality.

To reach certitude, there needs to be sufficient reasoning that he referred to as informal inference. This is a personal process of reasoning that highlights the fundamental distinction between inference and assent. Again, inference is conditional in the sense of being dependent upon the preceding inferences, and assent is unconditional in the sense of being independent of the preceding inferences. The meaning of sufficient reasoning is this: informal inferences justify certitude in religion or morality not by the abstract process of discursive or logical reasoning but by a personal process of discerning converging probabilities. This stance does not mean that logic or formal inference is not important. From early in his career, Newman insisted that formal inference provides an indispensable support for and check on informal inference for consistency and coherence. The contribution of discursive reasoning is "to analyze, verify, methodize, and exhibit" in a "process of analyzing and describing … which takes place upon reflection."[26] Yet, he was very wary of the limited reach of logic in matters of religious belief and morality. In a famous passage he voiced his concern with flair:

> Knowledge of premises, and inferences upon them,– this is not to *live*. It is very well a matter of liberal curiosity and of philosophy to analyze our modes of thought: but let this come second, … But if we commence with … argumentative proof, … or attempt to make man moral or religious by libraries and museums, let us in consistency take chemists for our cooks, and mineralogists for our masons.[27]

The personal reasoning process means that converging probabilities cannot compel the conclusion in the way that logic can. Nonetheless, personal reasoning can ascertain when accumulating probabilities touch upon a conclusion to justify it. This is a personal type of discernment that Newman described as moral demonstration

---

[24] Newman (1870a), 65–65, 390.

[25] Newman (1870a), 259.

[26] Newman (1843), 256, 263.

[27] Newman (1870a), 95–96.

in the sense of being virtually proven.[28] Probable arguments are sufficient when united together to reasonably justify the conclusion.

A variety of analogies can illustrate his point. One analogy is the comparison between a cable and an iron rod: a rod with its rigidity depicts the strict demonstration of formal inference or logic; the strands of a cable combine flexibly (akin to probabilities that cumulate together) to bear the weight of a particular conclusion–this represents the moral demonstration of informal inference.[29] Another analogy for moral demonstration is dovetailing that occurs in woodwork: a proposition is justified as being true when the conclusion dovetails so closely with preceding knowledge that nothing else but the conclusion fits.[30] Recalling the paradigm of marriage can be helpful here, though Newman never made this association. A couple reaches the point when there is sufficient reason to marry, with the decision justified by the informal inferences based upon the couple's experiences dovetailing together, even though the decision may not be capable of being proven by strict logic. The decision to wed is justified as being conditional upon the preceding inferences, yet sufficient to elicit the unconditional assent of certitude. In such a scenario, if certitude is not justified for the couple one could only smile at the consequences of declaring their love merely as a probability!

Hence, a shift occurs from affirming a conclusion as dependent upon the inferences to affirming the conclusion as independent from the inferences in certitude. The shift occurs because the conclusion is affirmed in its own right as objectively true. Certitude refers to the unconditional acceptance of a conclusion as true.[31] An important distinction is drawn here. The logical manipulation of propositions can elicit truth as being objectively demonstrative through the abstract reasoning of formal inference (using the analogy of the iron rod). In contrast, the personal process of informal inference (using the analogy of the flexible cable) can justify the assent of certitude. Here the conclusion is asserted independently of the preceding inferences, yet the independence of assent remains aligned with the subjective process of inference. There is no subject-free objectivity in this process! There is no dichotomy between the objective truth and the subjective affirmation of it in this process of imaginative discernment.

For Newman, certitude meant that a proposition is affirmed as true without reserve or doubt. However, this is different from the exclusion of doubt that occurs in the abstract process of discursive reason. In the formal inference of logic, doubt is excluded because the conclusion has been proven demonstratively in the sense that the abstract inferences compel the conclusion. There is no doubt because the conclusion is logically proven. In contrast, in the personal process of informal inference, the conclusion is only virtually proven in the sense that the conclusion cannot be otherwise and it would be irrational to reject it.

---

[28] Newman (1843), 274.

[29] Newman (1864b), 126.

[30] Newman (1870a), 327.

[31] Newman (1870a), 329, 411.

> We decide, not that the conclusion must be, but that it cannot be otherwise. We say, that we do not see our way to doubt it, that it is impossible to doubt, that we are bound to believe it, that we should be idiots, if we did not believe.[32]

Only in this sense is doubt about the conclusion excluded. Unlike formal inference, in the assent of certitude there remains a logical distinction between the informal inferences and the justified conclusion. That distinction means that doubt might be theoretically possible, even though it is not practically feasible given the justification of the conclusion. Newman helpfully illustrated his point on doubt by alluding to the judgment of a jury: the evidence necessary for a verdict, even if lacking discursive proof, can be deemed sufficient based on converging inferences to judge a prisoner guilty insofar as there is no reasonable doubt, though theoretical or logical doubt remains possible.[33] In other words, when a conclusion is virtually proven to justify certitude, there is no practical reason to doubt it. The concrete process of inference and assent is subjective, involving moral evidence, moral demonstration, and moral certitude:

> This certitude and this evidence are often called moral; ... I observe that moral evidence and moral certitude are all that we can attain, ... in the case of ethical and spiritual subjects[34] (*GA*, 318).

By using the word "moral" he did not mean practical certainty as merely safe to act upon. For Newman probable arguments can justify a conclusion to be held with certitude; in contrast, conclusions that are merely safe to act upon remain at the level of probability. His epistemology was designed to shift from probability to certitude. He used the word "moral" to refer to the subjective nature of converging probabilities that justify certitude as objectively true. This dynamic interaction between subjectivity and objectivity was normative for his religious belief and morality. He was very aware of the difference between his own approach and the traditional argument of Aristotle's *phronesis*. Indeed, he compared his concrete process of informal inference with Aristotle's understanding of practical judgment, *phronesis*. But, it is important to emphasize that the comparison he made was between Aristotle's *phronesis* and converging probabilities as dependent upon the inferences (not between *phronesis* and the assent of certitude): "It is not with assent, but with the controlling principle in inferences that I am comparing *phronesis*."[35]

The role of the imagination in this shift from informal inference to certitude is crucial. The imagination provides images around which reasoning can coalesce relevant inferences in a persuasive manner to justify a conclusion. Imaginatively, the cords of the flexible cable weave together, just as inferences converge together for a jury in the courtroom, to justify a conclusion being held firmly, with certitude, beyond reasonable doubt. This is how we justify belief in God or the decision to

---

[32] Newman (1870a), 317.

[33] Newman (1870a), 325–326.

[34] Newman (1870a), 318; (1864b), 146

[35] Newman (1870a), 354–356.

marry. The image of a God or of being together in marriage enables relevant inferences to combine meaningfully towards certitude.

The imagination also has another function in Newman's religious epistemology, in the sense that the intensity of its images can inspire action. Newman was skeptical of the so-called armchair ethicist who advances conclusions based on deductive, abstract inferences that are detached from the concrete reality of personal experience–these conclusions all too often remain detached from accompanying action.

> And this again is the secret of the distrust and raillery with which moralists have been so commonly visited. They say and do not. Why? Because they are contemplating the fitness of things, and they live by the square, when they should be realizing their high maxims in the concrete.... I have no confidence, then, in philosophers who ... sit at home, and reach forward to distances which astonish us; but they hit without grasping, and are sometimes as confident about shadows as realities.[36]

Newman recognized that the imagination not only could provide experiential images for informal inferences to converge toward a conclusion but that the imagination also provides an intensity that sparks action. The creative characteristics of the imagination have the capacity to stimulate action arising from certitude. The intensity and vividness of concrete images lead to action. In this sense, the imagination provides the evocative means to stimulate action. He pointed to the example of slavery to make this point. When the iniquity of the slave trade was held merely as a notional abstraction by discursive reasoning, action tended not to follow. He noted that the iniquity of the slave trade was one of the "great truths" accepted by society. But it was held merely as a notional abstraction having little impact upon practice: "the iniquity … of the slave-trade … needed an organized agitation … to affect the imagination of men as to make their acknowledgement of that iniquitousness operative."[37] When the imagination became involved in the concrete process of informal inference and certitude, the recognition of slavery as iniquitous became operative–action resulted. It is worth recalling that the Slavery Abolition Act in Britain occurred in 1833, well into the lifetime of Newman.

Newman's approach emphasized that the application of abstract moral law unavoidably varies for different individuals in different circumstances. He had little confidence in what he called "the dead letter of a treatise of a code."[38] Rather, he trusted the flexibility of informal inference as "a capacity sufficient for the occasion, deciding what ought to be done here and now, by this given person, under these given circumstances."[39] Informal inference is able to apply general principles to moral practice:

---

[36] Newman (1870a), 93.

[37] Newman (1870a), 77.

[38] Newman (1870a), 356.

[39] Newman (1870a), 355.

> It is to the living mind that we must look for the means of using correctly principles of whatever kind, ... and of discerning what conclusion from these is necessary, suitable, or expedient.[40]

Even if there is a high probability that an abstract moral law pertains deductively in a specific case, it is the reasoning of informal inference that elicits an assent of certitude. In his *Apologia Pro Vita Sua* he exemplified this shift from a "dead" abstract code to the "living" discernment of moral action by referring to the ethical law against lying.[41] Although the moral law against lying was objectively true as an abstract notion, there can be a justification of equivocation in the specific circumstances surrounding the controversy. That is, equivocation can be justified in practice despite the moral law against lying. Subsequently, that distinction became much more famous at the time of World War II when householders lied to authorities to shield victims from arrest and internment camps.

For Newman there was a crucial distinction between notional recognition of abstract truth and the evocative recognition of the same truth in the concrete process of informal inference. That is why he excoriated ethicists for pontificating moral truths without accompanying action, that is, for not implementing abstract maxims in concrete practice. Abstract deductions, he argued, have little power to inspire action, leading to his famous remark that contrasted the perception of religious dogma in certitude with abstract inferences of deductive logic: "The heart is commonly reached, not through the reason, but through the imagination, … Many a man will live and die upon a dogma: no man will be a martyr for a conclusion."[42] This dissonance between abstract and concrete reasoning shaped his theological method where he prioritized the personal approach of informal inference and certitude that would connect conclusion drawn with actions taken: "It is otherwise with the theology of a religious imagination. It has a living hold on truths."[43]

The distinction between abstract and concrete inferences did not exclude a legitimate place for both types of reasoning in theology. His theological method can be described as integrating both the abstract reasoning process of formal inference and the concrete reasoning process of informal inference that elicits certitude. By distinguishing between notional and real apprehension of truth, he did not mean to polarize them as being inconsistent with each other. Rather, the same proposition can express what is notional and what is real in the same mind at the same time, though the primary focus must be on what is concrete and real. For example, he explained that a dogma is a proposition that can elicit an abstract assent by the theological intellect and a concrete or real assent by the religious imagination.[44] Again, he was not polarizing the theological intellect and the religious imagination, because each needs the other. Rather, he used these different terms (theological intellect and

---

[40] Newman (1870a), 360–361.

[41] Newman (1864a).

[42] Newman (1870a), 92–93.

[43] Newman (1870a), 117.

[44] Newman (1870a), 10–11.

religious imagination) as a rhetorical device to distinguish the abstract and concrete reasoning processes in his theological method that he described as the theology of a religious imagination. Newman's hermeneutics of the imagination enabled him to integrate his religious epistemology with his theological method to present surprisingly critical views on the tension between personal conscience and Church tradition, a tension that continues today to haunt debates on moral normativity. This distinctive aspect of this view of normativity in religious morality is explored in the next section.

5.4.2 *Personal Conscience & Church Tradition*

Newman is well known for his writings on conscience. However, without interpreting it in light of his religious epistemology as discussed above, his contribution can be misunderstood. He famously described conscience as combining a moral sense and a sense of duty:

> Conscience has a legitimate place among our mental acts ... it is a moral sense, and a sense of duty; a judgment of the reason and a magisterial dictate. Of course its act is indivisible; still it has these two aspects, distinct from each other, and admitting of a separate consideration.... Thus conscience has both a critical and a judicial office, ...its testimony that there is a right and a wrong, and its sanction to that testimony conveyed in the feelings which attend on the right or wrong conduct. Here I have to speak of conscience ... as supplying us, by means of its various acts, with the elements of morals, such as may be developed by the intellect into an ethical code, ... (and) as the dictate of an authoritative monitor bearing upon the details of conduct as they come before us, ... as a rule of right conduct, ... as a sanction of right conduct.[45]

Typically, this distinction is interpreted as meaning the moral sense deals with moral laws and the sense of duty indicates a religious sanction as a voice of God, so to speak. There is good reason for such a stance based on Newman's own words, presenting "the phenomena of Conscience, as a dictate, ... with a picture of a Supreme Governor, a Judge, holy, just, powerful, all-seeing, retributive, ... as the Moral Sense is the principle of ethics."[46] However, such an interpretation separates his view of conscience from his basic religious epistemology in this sense. On the one hand, while respecting a legitimate place for moral law, his epistemology privileges the role of informal inference in implementing abstract laws in daily life. His epistemology gives priority to informal inference, not the abstract formulations of formal inference in moral laws. In other words, the moral sense of conscience needs to be interpreted in a manner that is consistent with the dynamic between formal and informal inference in his epistemology. On the other hand, it would appear odd to consider the sense of duty of conscience merely as a voice of God insofar as he presented the above description of "a Supreme Governor, a Judge" as characterizing

[45] Newman (1870a), 105–106.

[46] Newman (1870a), 110.

the unreflective assent of a child: "It is the image of One who is good, inasmuch as enjoining and enforcing what is right and good, and who, in consequence, not only excites in the child hope and fear, ... but kindles in him love towards Him, as giving him a good law."[47] In other words, the sense of duty needs to be interpreted in a manner that is consistent with the critical assent of certitude (in contrast to the uncritical assent of children).

A deeper interpretation of conscience as a moral sense and a sense of duty can be presented in light of his sophisticated religious epistemology. This interpretation needs to integrate these related aspects of conscience with the priority of informal inference in his epistemology. This can be done by an autonomous and a theonomous interpretation of conscience. First, the moral sense can be construed as representing the autonomous function of conscience that combines abstract formal inference with personal informal inference with regard to morality. This means that the rational discernment of conscience in morality should prioritize personal reasoning over abstract reasoning, informal inference over formal inference, to apply abstract moral law in concrete historical reality. The implications of this approach were mentioned at the start of this essay when discussing practical moral dilemmas like the Phoenix case. Second, the sense of duty, construed as representing the theonomous function of conscience, also needs to be consistent with his epistemology that combines abstract formal inference with personal informal inference. Here, a more nuanced explanation is needed than merely associating the sense of duty with divine sanction as exemplified in the mind of a child.

Newman's mature epistemology provides a subtler interpretation of the sense of duty. He adopted a traditional view of Natural Law as "the voice of God in the nature and heart of man, as distinct from the voice of Revelation",[48] referring to "truths which the Lawmaker has sown in our very nature."[49] By seeking to interpret the voice of God through nature as a rational endeavor, the sense of duty can be construed as being similar to the moral sense. The distinctive characteristic of the sense of duty is that it has a theonomous character (regarding the voice of God) in contrast to the autonomous character of the sense of duty (focusing on reasoning alone). This theonomous character enables the rational process of moral discernment to meaningfully depict secular perceptions in religious language. For example, discerning that X is wrong (as a function of the moral sense) can be depicted as X being a sin (as a function of the sense of duty). In other words, rational discernment in the sense of duty does not inherently change from the epistemological process of the moral sense just because we seek to trace moral reasoning to a transcendent source (God). A similar approach to conscience was developed by the renowned German moral theologian Josef Fuchs (without reference to Newman) in the twentieth century. Fuchs referred to the theonomy of moral autonomy to explain that "the moral dimension, by remaining totally itself, is penetrated, enriched and fulfilled in

[47] Newman (1870a), 113–114.
[48] Newman (1896), 247.
[49] Newman (1896), 253.

every respect by the theological-religious dimension which grants it a meaning that it did not properly possess."[50]

It would be mistaken to construe this theonomous explanation of the autonomy of conscience as referring simply to motivation. Indeed, a religious motive is involved, that is, doing something in the name of God. However, this view of the sense of duty implies a profound transformation of the rational discernment of the moral sense. Again, the example of marriage can illustrate this point. When a couple is married, the religious significance of their marriage is transformed from the relation that the couple had prior to being wed, a new significance that can be articulated in terms of sacramental meaning with regard to holiness and salvation. In other words, for Newman the autonomous function of conscience in the informal inference process of the moral sense can elicit a theonomous interpretation in the sense of duty that provides religious meaning to rational moral discernment.

Interpreting Newman's conscience in this manner helps to clarify his approach to Church tradition, especially regarding his support for episcopal authority. It is well known that his view of Church tradition celebrated the reciprocity between Bishops, theologians, and the faithful as indispensable for discerning truth both in religious belief and in moral discernment. In his essay *On Consulting the Faithful in Matters of Doctrine* he introduced the concept of an ecclesiastical sense as a type of discernment in the Church. He associated this ecclesiastical sense with the reciprocity between Bishops, theologians, the faithful, using three Latin terms: the *ecclesia docens* (translated as the teaching Church, referring to the teaching authority of the Bishops), the *schola theologorum* (translated as schools of theology, hence referring to theologians), and the *consensus fidelium* (translated as the consent or assent of the laity and religious in the Church, referred to as the faithful): "Each constituent portion of the Church has its proper functions, and no portion can safely be neglected."[51]

He consistently upheld the authority of the bishops as being the official teaching office of the Church: "the gift of discerning, discriminating, defining, promulgating, and enforcing any portion of that tradition resides solely in the Ecclesia docens."[52] However, their authority was not isolated. He insisted upon the need to consult the faithful in matters of doctrine: "the fidelium sensus and consensus is a branch of evidence which it is natural or necessary for the Church to regard and consult, before she proceeds to any definition."[53] And he connected the sense of the faithful closely with the role of theologians: "For myself I think the securus judicat orbis terrarum, is the real rule and interpretation of the words of the Church, i.e. the sensus theologorum primarily, then consensus fidelium next."[54] By the latin phrase "securus judicat orbis terrarium" he meant "the deliberate judgment, in which the

[50] Fuchs (1983), 101.

[51] Newman (1859), 103.

[52] Newman (1859), 63.

[53] Newman (1859), 55.

[54] Newman (1867), 254.

whole Church at length rests and acquiesces."[55] The basic point that he sought to make was that theology has an indispensable role in the discernment of truth in Church tradition. Late in life, in his 1877 preface to the *Via Media* (written in 1839), he wrote: "Theology is the fundamental and regulating principle of the whole Church system."[56] Hence, he ardently defended freedom in theology: "The freedom of the Schools, indeed, is one of those rights of reason, which the Church is too wise really to interfere with. And this applies not to moral questions only, but to dogmatic too."[57] As mentioned above, by theology he meant both the abstract reasoning of formal inference and the concrete reasoning of informal inference, with the latter having priority of importance. Also, when referring to theology he specifically included questions or morality, underscoring the importance of theology in discussion of moral normativity.

Moreover, he was cognizant that dissent can be a significant part of religious assent in theological discourse. After all, dissent had accompanied him throughout his life: he had dissented from the stance of the Anglican Bishops in 1841 over *Tract 90* that he had written in a widely distributed series of tracts where he initiated a fiery controversy over a middle way (*via media*) between Anglicanism and Catholicism. He explained: "I yielded to the Bishops in outward act, viz. in not defending the Tract, ... not only did I not assent inwardly to any condemnation of it, but I opposed myself to the proposition of a condemnation on the party of authority."[58] A few years later, in 1845 when he converted to Catholicism he dissented from Anglicanism in a soul-deep manner: "My sole ascertainable reason for moving is a feeling of indefinite *risk* to my soul in staying ... I don't think I *could* die in our communion."[59] Over a decade later, as a Catholic priest, he objected when the Catholic Bishops opposed his essay *On Consulting the Faithful* in 1859 (published as an essay in the *Rambler*, a Catholic journal that he edited) causing him to refrain from writing for several years: "The cause of my not writing from 1859 to 1864 was my failure with the *Rambler*. I thought I had got into a scrap and it became me to be silent."[60] In that silence he did not withdraw his essay, or the views in it, on the faithful. Referring to these traumatic personal experiences of being publicly criticized by Bishops, he wryly remarked that a theologian "would be fighting, as the Persian soldiers, under the lash, and the freedom of his (sic) intellect might truly be said to be beaten out of him."[61]

Even though he defended the authority of the Bishops as the *ecclesia docens*, it can be surprising for readers today to learn of his profound skepticism regarding discussions of Papal Infallibility at Vatican I. He clearly dissented antecedently,

[55] Newman (1864a), 110.
[56] Newman (1877), xlvii.
[57] Newman (1864a): 447.
[58] Newman (1864a), 416.
[59] Newman (1844), 427.
[60] Tristram (1957), 272.
[61] Newman (1864a), 239.

explaining that he could not believe Papal Infallibility as a dogma, while acknowledging that he would abide by it if passed, as eventually transpired. The specifics of his opposition to the doctrine of Papal Infallibility are gripping, revealing an artful capacity for theological nuance. From his days as an Anglican, he had recognized the Infallibility of the Church, a recognition that continued after becoming a Catholic in 1845. In his *Apologia Pro Vita Sua* he explained that the Infallibility of the Church resides with the Pope in Ecumenical Council, guided by scripture and tradition. However, he was very worried about a declaration of Papal Infallibility occurring independently of a Council–that being precisely what Vatican I anticipated for future Popes. Newman perceived the Church as being led astray by Ultramontanism, a conservative movement in the Catholic Church that he considered to be fanatical. Also, he was disturbed that there was not an effective process to consult the faithful on Papal Infallibility as a new Church doctrine. He had strenuously advocated in his 1859 essay that the faithful should be consulted antecedently in such matters of doctrine. Hence, he wondered in 1870 what the faithful (including himself) had done to be treated in such a dismissive fashion over the debate on Papal Infallibility: "What have we done to be treated, as the faithful were never treated before."[62] Also, he worried about the negative impact the doctrine might have on educated Catholic laity. After the doctrine was defined, he wrote anxiously in a letter: "the actual tendency of the definition then in prospect will be to create in educated Catholics a habit of skepticism or secret infidelity as regards all dogmatic truth."[63] Newman anticipated just after Vatican I that doctrinal development around Papal Infallibility would occur:

> The late definition does not so much need to be undone, as to be completed. It needs *safeguards* to the Pope's possible acts–explanations as to the matter and extent of his power…. Let us be patient, let us have faith, and a new Pope, and a re-assembled Council may trim the boat.[64]

Though such development has not yet happened, it is worth noting that no Pope has applied this doctrine to matters of morality. Perhaps the fears that Newman articulated about the doctrine may have caused the Church subsequently to be extraordinarily cautious about applying the doctrine to specific topics.

Because Newman deemed the 1870 declaration to be more moderate than many had anticipated, he eventually accepted the doctrine. But he was very cautious in doing so: he waited until the Bishops who had left the Council early (related with the strife surrounding Rome as Italy became a nation) had voiced support so that the doctrine could robustly claim the authority of an Ecumenical Council. But he could not let go of the controversy. Just a few years later, he appeared to gleefully take the opportunity of an after dinner toast to balance the authority of the Pope with that of conscience. The remark appears in his 1874 *Letter to the Duke of Norfolk*:

---

[62] Newman (1870b), 18.

[63] Newman (1870b), 166.

[64] Newman (1871), 310.

> Certainly, if I am obliged to bring religion into after-dinner toasts, (which indeed does not seem quite the thing) I shall drink–to the Pope, if you please,–still, to Conscience first, and to the Pope afterwards.[65]

To comprehend this bold remark, it is useful to note that Newman was defending what he referred to as "the supremacy of Conscience" versus "an absolute obedience" to the Pope in the event of future situations where a Pope might be "transgressing the laws of human society."[66] The point appears to be as follows. Insofar as the primary function of conscience is the rational judgment of the moral sense, if a case ever arose where a Pope was found to be transgressing justified laws of human society, then individual conscience could legitimately reject that transgression. That rejection would be based on the authority of the moral sense of conscience. Hence, Newman's toast to conscience seems to reflect his commitment to the truth that the moral sense perceives through the concrete process of informal inference, that is, imaginative discernment.

5.5 Conclusion

At the core of Newman's view of moral normativity was the critical role of the personal process of informal inference that justifies certitude. He recognized the legitimacy of the abstract reasoning process of formal inference, but argued that it was a secondary though necessary mode of reasoning in matters of religious belief and morality. It was the crucial role of informal inference that guided both his understanding of individual conscience and his view of Church tradition, leading him to be cautious about the abstractions of moral laws. Moral normativity must be accountable primarily to the process of informal inference that guides personal conscience, and not merely to abstract moral laws even when they carry the authority of Church tradition. Newman sought to address the tension between personal conscience and Church tradition by turning to theology, relying on a theological method that implemented his religious epistemology. His theology of a religious imagination makes a distinctive contribution by highlighting a place for the imagination in moral normativity. The analysis suggests that the abstract process of formal reasoning adopted in the classical principle of double effect can be enhanced via this role of the imagination. The role facilitates personal and practical reasoning to coalesce towards insight and moral conclusions that can subsequently be explained using double effect reasoning. This approach helps to foster doctrinal development in Catholic bioethics.

The role of the imagination is not to supplant rationality: the imagination does not provide a mode of reasoning. Rather, the imagination has two constructive purposes in moral normativity. First, the imagination can provide riveting images based

[65] Newman (1874), 261.

[66] Newman (1874), 243.

on personal experience that enable reasons to coalesce in a meaningful manner around a conclusion. When that occurs, the primary role is played by personal, experiential reasoning; but the insights that accrue in this process must be held accountable to the abstract processes of formal inference. In other words, for Newman, both modes of reasoning function in a reciprocally supportive manner. The examples of marriage or belief in God illustrate this interaction. Second, the images of the imagination can generate an intensity and vividness of concrete experiences that lead to action. Many examples illustrate this dynamism of imaginative discernment: the image of urgently saving the life that can be saved, such as in the Phoenix case; the image of God used by Pope John Paul II to prohibit the death penalty; and the image of micropcephaly that Pope Francis used to legitimize using contraception to prevent newborns being damaged by Zika virus infection. Hopefully, we can become more attuned to the experiential capacity of imaginative discernment, integrating formal and informal reasoning, to yield new insights in morality.

Suggested Readings

Banner, M. 2016. *Ethics of Everyday Life: Moral Theology, Social Anthropology, and the Imagination of the Human*. New York: Oxford University Press.

Greeley, Andrew. 2001. *The Catholic Imagination*. Berkeley: University of California Press.

Jennings, W.J. 2011. *The Christian Imagination*. New Haven: Yale University Press.

Kaag, J. 2014. *Thinking Through the Imagination. Aesthetics in Human Cognition*. New York: Fordham University Press.

Kieran, M., and Lopes, D., eds. 2012. *Imagination, Philosophy and the Arts*. London: Routledge.

Kind, A. 2016. *The Routledge Handbook of Philosophy of Imagination*. London: Routledge.

Magill, G. 2015. *The Religious Morality of John Henry Newman. Hermeneutics of the Imagination*. New York: Springer.

References

Abelson, R., et al. 1977. *The Philosophical Imagination. An Introduction to Philosophy*. New York: St. Martin's Press.

Catechism. 2000. *Catechism of the Catholic Church*. Rome: Vatican City.

Code of Canon Law. 1983. Rome: Vatican City.

Fuchs, Josef. 1983. *Personal Responsibility and Christian Morality*. Dublin: Gill and Macmillan.

Gardner, H. 1982. *In Defense of the Imagination*. Oxford: Clarendon Press.

Janssens, J. 1972. Ontic Evil and Moral Evil. *Louvain Studies* 4: 115–156.

Johnson, M. 1987. The Body in the Mind. In *The Bodily Basis of Meaning, Imagination, and Reason*. Chicago: University of Chicago Press.
———. 1993. *Moral Imagination. Implications of Cognitive Science for Ethics*. Chicago: University of Chicago Press.
Kaczor, C. 1998. Double-Effect Reasoning From Jean Pierre Gury to Peter Knauer. *Theological Studies* 59: 297–315.
Kaufman, G. 1981. *The Theological Imagination. Constructing the Concept of God*. Philadelphia: Westminster Press.
Keenan, J.F. 1993. The Function of the Principle of Double Effect. *Theological Studies* 54: 294–315.
Kelly, D.F., G. Magill, and H. ten Have. 2013. *Contemporary Catholic Health Care Ethics*. Washington, DC: Georgetown University Press.
Knauer, P. 1967. The Hermeneutic Function of the Principle of Double Effect. *Natural Law Forum* 12: 132–162.
Liebard, O.M. 1978. *Love and Sexuality, Official Catholic Teachings*. Wilmington: Consortium.
Lovibond, S. 1983. *Realism and Imagination in Ethics*. Oxford: Blackwell.
Lysaught, M.T. 2011. Moral Analysis of Procedure at Phoenix Hospital. *Origins* 40: 537–549, at 538.
Mackey, J.P., ed. 1986. *The Religious Imagination*. Edinburgh: Edinburgh University Press.
Magill, G. 2011. Threat of Imminent Death in Pregnancy: A Role for Double effect Reasoning. *Theological Studies* 72: 848–878.
Mangan, J.T. 1949. An Historical Analysis of the Principle of Double Effect. *Theological Studies* 10: 41–61.
McIntyre, J. 1987. *Faith, Theology and Imagination*. Edinburgh: The Handsel Press.
Newman, J.H. 1843. *Newman's University Sermons: Fifteen Sermons Preached before the University of Oxford 1826–43*. Edited and Introduction by D.M. McKinnon, and J.D. Holmes. London: SPCK, 1970.
———. 1844. *The Letters and Diaries of John Henry Newman*. Vol. X. Oxford: Oxford University Press, 2006.
———. 1859. *On Consulting the Faithful in Matters of Doctrine*. Edited with an Introduction by J. Coulson. London: Geoffrey Chapman, 1961.
———. 1864a. *Apologia Pro Vita Sua*. Edited, Introduction and Notes by M. J. Svaglic. Oxford: Clarendon Press 1967.
———. 1864b. *The Letters and Diaries of John Henry Newman*. Vol. XXI. Oxford: Oxford University Press, 1971.
———. 1867. *The Letters and Diaries of John Henry Newman*. Vol. XXIII. Oxford: Oxford University Press, 1973.
———. 1870a. *An Essay in Aid of a Grammar of Assent*. Edited, Introduction, and Notes by I. Ker. Oxford: Clarendon Press, 1985.
———. 1870b. *The Letters and Diaries of John Henry Newman*. Vol. XXV. Oxford: Oxford University Press, 1973.
———. 1871. *The Letters and Diaries of John Henry Newman*. Vol. XXV. Oxford: Oxford University Press, 1973.
———. 1874. Letter to the Duke of Norfolk. In *Certain Difficulties felt by Anglicans in Catholic Teaching*, vol. II. London: Longmans, Green & Co., 1896.
———. 1877. *The Via Media of the Anglican Church*. Vol. I. In, Edited with Introduction and Notes by H.D. Weidner. Oxford: Clarendon Press.
———. 1896. *Certain Difficulties felt by Anglicans in Catholic Teaching*. Vol. II. London: Longmans, Green & Co.
Pope John Paul II. 1993. *The Splendor of Truth*. Rome: Vatican City.
———. 1994. *Apostolic Letter, Ordinatio Sacerdotalis*. Rome: Vatican City.
———. 1995. *The Gospel of Life*. Rome: Vatican City.
Pope Paul VI. 1968. *Humanae Vitae*. Rome: Vatican City.

Pope Pius XII. 1951. Address to the Associations of the Large Families. *Acta Apostolicae Sedis* 43 (November 26): 855–860.
Ricoeur, P. 1995. *Figuring the Sacred. Religion, Narrative, and Imagination*. Translated by David Pellauer and edited by Mark. I. Wallace. Minneapolis: Fortress Press.
Touhey, J.F. 2011. "A Fatal Conflict: Can Catholic Hospitals Refuse to Save Lives?" *Commonweal* 38:2 (January 28): 8–10.
Tracy, D. 1981. *The Analogical Imagination. Christian Theology and the Culture of Pluralism*. London: SCM.
Tristram, H., ed. 1957. *John Henry Newman. Autobiographical Writings*. Edited with an Introduction by H. Tristram. London: Sheed and Ward.
U.S. Bishops. 2009. *Ethical and Religious Directives for Catholic Health Care Services*. 5th ed. USCCB: Washington, DC.
———. 2010. Committee on Doctrine. "The Distinction Between Direct Abortion and Legitimate Medical Procedures." USCCB, Washington, DC, June 23.
Warnock, M. 1976. *Imagination*. London: Faber and Faber.

Chapter 6
Normative Bioethics in Hinduism

Joris Gielen

Abstract When the possibility of normative bioethics is assessed against the broader background of Hinduism, it becomes clear that normativity is not an obvious concept in this religion. Difficulties related to defining Hinduism make it equally hard to describe essential characteristics of Hindu bioethics. These problems are further compounded by the very limited number of publications on Hindu bioethics. Therefore, this study largely relies on empirical data to understand how Hindus develop normative arguments on bioethical issues. This investigation shows that Hindus apply religious concepts and ideas to bioethical issues through an approach that combines virtue ethics and deontological ethics. In the empirical data, karma appears as a guiding belief for arguments that are in line with virtue ethics, while belief in an almighty God motivates deontological arguments. Hindus themselves do not make a clear distinction between these two approaches, but rather combine them in an overarching argument. This Hindu way of dealing with bioethical issues can be understood through the philosophical concept *saṃskāra*. Through this approach, Hindu bioethics is able to integrate various perspectives and avoids the pitfalls of both ethical absolutism and relativism. This makes Hindu bioethics a very relevant example when scholars attempt to find ways to deal with normativity in global bioethics.

Keywords Hinduism · Bioethics · Normativity · Virtue ethics · Deontological ethics · *Saṃskāra* · Ethical absolutism · Ethical normativity

J. Gielen (✉)
Center for Healthcare Ethics, Duquesne University, Pittsburgh, PA, USA
e-mail: gielenj@duq.edu

J. Gielen (ed.), *Dealing with Bioethical Issues in a Globalized World*,
Advancing Global Bioethics 14, https://doi.org/10.1007/978-3-030-30432-4_6

6.1 Introduction

India is known as a country of contrasts that can be perceived in all areas of human life, such as religion, the economy, and even healthcare. India's cultural and religious diversity often amazes travelers who can see a wide range of religious practices and hear about diverging religious beliefs. While the economic liberalization since the nineties has led to a burgeoning middle class that has adopted a consumerist lifestyle, 21.9% of the population still lives below the country's national poverty line (UNESCAP-SSWA 2015). In India, healthcare, too, is characterized by diversity. Besides Western medicine, traditional systems of medicine such as Ayurveda and Unani but also homeopathy enjoy wide popularity (Singh et al. 2005). The popularity of those systems of medicine can be partially explained by the fact that good biomedical facilities are not easily available in many parts of India. Notwithstanding such shortcomings in coverage, over the last two decades, healthcare based on Western biomedical science has been has been rapidly expanding (Gudwani et al. 2012). Advanced medical treatment is available in Indian cities and the country has even developed into a hotbed of medical tourism and research.

When we pull all the above mentioned aspects together we can easily understand how, in India, bioethical issues abound. There are the issues of advanced medical treatment, which also arise in Western contexts, such as ethical issues related to genetics or forgoing treatment at the end of life. Besides these, there are issues related to equity and access to healthcare. In medical research, the risk of exploitation of poor and vulnerable research subjects looms large. These are just a few examples that illustrate the ubiquity of bioethical issues in India. When people in India search for solutions to these issues, their religion might guide their answers. Religion may direct them towards a normative conclusion regarding the ethical acceptability of research, treatments and procedures in biomedicine.

In this chapter, we will have a look at how normative bioethical arguments are developed in Hinduism, the majority religion in India. To do this, I will first situate Hindu bioethics within the broader context of the Hindu religious tradition. I will, then, explain the particular nature of normative bioethics in Hinduism by comparing it to Western ethical theories. To do this, I will rely on empirical data that I have collected in healthcare in India over the past decade.

6.2 Defining Hinduism and Hindu Ethics

If we want to understand the nature of Hindu bioethics and the role of normativity within it, it would be helpful if we could first outline the essential characteristics of Hinduism in general, because we can expect that these characteristics will also surface in Hindu bioethics. We could even argue that bioethics can only be rightly called "Hindu" if it incorporates these essential characteristics of Hinduism. As justifiable and straightforward as this assertion may sound, it does pose serious challenges, because an analysis of the phenomena that are commonly referred to as "Hindu" reveals that such characteristics are elusive. This becomes particularly clear when we try to define what makes a person a Hindu. In India, millions of people self-identify as Hindu. In the 2011 census, 966,257,353 people or 97.8% of the entire population were listed as Hindu (India 2016). But, is there anything that all these people share beyond the common nametag of being Hindu? If we were to undertake a survey among them, we would observe that common or essential religious characteristics for all of them are hard to find. Aspects that could be used as identifiers of belonging to a religion, such as reverence for a religion's prophetic founder or founding event, acceptance of Holy Scripture, or belief in particular dogmas, do not have the same capacity of religious benchmarks in Hinduism as they do have in other religions.

First, in Hinduism, there is no prophetic founder or founding event that started this religion. Hindus themselves have often described Hinduism as *sanātana dharma,* which can be translated as "the eternal religion." Hindus who define their religion in this way see Hinduism as a particular search for religious and philosophical truth or a body of devotional beliefs and practices that has been around forever. From this perspective, there is not one point in history that can be earmarked as the beginning of Hinduism. Second, in Hinduism, there is no Holy Scripture that is central in the lives of all Hindus and serves as a source of inspiration for the decisions they make in life. In Hinduism, there rather is a plethora of religious texts, which often contain contradictory views and whose use and centrality differ among Hindu groups. At first, the *Veda* may seem an exception to this rule, because this collection of ancient scriptures has an important place in the religious life of a majority of Hindus. Nevertheless, its stature markedly differs from that of the Bible in Christianity, the Torah in Judaism, or the Quran in Islam. The Veda is almost exclusively recited in a ritual context in which the sound of the words is more important than their meaning (Balagangadhara 1994, pp. 416–418; Staal 1996, pp. 201–208). Moreover, many devotional practices and religious views that are central to many Hindus today are not even mentioned in the Veda. The Veda is not the direct source of most of Hinduism's religious beliefs and practices. Third, in Hinduism, there are no dogmas or essential tenets that should be accepted by every Hindu. There are, indeed, several convictions and lines of conduct to which many Hindus adhere, but these do not have the same consequences of articles of faith, such as belief in the Trinity in Christianity. Therefore, it does not make much sense to attempt to determine whether or not persons are Hindus on the basis of their acceptance of certain religious convictions or lines of conduct.

In this regard, belief in karma constitutes an interesting case. We will discuss the bioethical implications of this belief later in this chapter. Karma implies the conviction that every deed will have a consequence in this life or in a future one. Good deeds will have good consequences, while evil deeds will have bad consequences. Almost without exception, Hindus will accept karma as an important aspect of their worldview. Yet, can we conclude from there that karma has the status of dogma in Hinduism? That seems to be one bridge too far. Karma, in essence, is just a word or concept that is in need of specific elaboration in a religious context. As we will see below, within Hinduism there are diverging theories about how karma works and how it affects the lives of living beings. Moreover, the concept figures prominently in other religions that originated in India, namely Buddhism, Sikhism, and Jainism, and, as such, it cannot function as a particular distinguishing characteristic of Hinduism.

These observations show how difficult it is to set apart essential characteristics of Hinduism. Considering these difficulties related to defining Hinduism some scholars, such as S.N. Balagangadhara (1994), Heinrich von Stietencron (2007), and Frits Staal (1996, pp. 387–419), have argued that 'Hinduism' is an artificial construction created by colonial explorers, officials, and academics who mistakenly projected Western notions of religion onto Indian culture and society. We do not need to go into the details of their arguments here, but we do have to be aware that their observations could have a decisive impact on the topic of the current study: normativity in Hindu bioethics. For, if Hinduism is an artificial construction, automatically, Hindu bioethics, which has to be considered a part of Hinduism, cannot be real either and any claim of normativity in a purported Hindu bioethical discourse must be judged hollow.

In this chapter, we will not follow that reasoning. We should not forget that, nowadays, all over the world, many people self-identify as Hindu. These claims of belonging to Hinduism do, in no way, become less relevant by arguments that show the involvement of western colonials and academics in the origination of the Hindu religious identity (Gielen 2013, pp. 37–39; Pennington 2005, p. 5). For many people who self-identify as Hindu, Hinduism may offer guidance or, we could even say, normative directions in their lives. This may be particularly true in healthcare and biomedical research, where difficult ethical issues abound.

Starting from the assumption that Hinduism could, indeed, offer normative answers to bioethical issues, a few authors have either tried to construct a bioethical model based on Hindu sources and ideas (e.g. Crawford 1995, 2003) or attempted to determine what Hindus who logically and systematically adhere to the basic principles and beliefs of their religion ought to think or can be expected to think about the treatment decisions (e.g. Coward and Sidhu 2000; Deshpande et al. 2005). Katherine K. Young (1989) studied normative Hindu views from a historical point of view by attempting to retrieve the evolution in opinions on treatment decisions at the end of life. Young tried to describe how Hindus thought or were supposed to think about euthanasia and related matters in the past, as this may help to understand attitudes in the present.

As interesting as the models and arguments developed by these authors may be, it is hard to tell to what degree or in what sense these publications can claim normative status in Hinduism. We can even wonder, more broadly, whether or not the arguments developed in them represent authentic Hindu positions. While evaluating publications such as the above mentioned studies on Hindu bioethics, Francis X. Clooney (1995, p. 440) remarked that these attempts to construct Hindu ethical positions "naturally privilege existent forms of ethical discourse in the West, gleaning from the chosen non-Christian sources answers to questions already posed in the expected modes of Western ethics." These scholars, most of whom are not

Hindus, are "speaking for, and in the place of Hindus – who do not usually write in this fashion." This assessment, although by now over 20 years old, still holds true in the sense that the number of publications on bioethics authored by Hindus is comparatively limited, especially when we compare it to the abundance of bioethical literature in other religious traditions, most notably Roman Catholicism and Judaism. Should we conclude from this that bioethics in Hinduism is a marginal phenomenon or hardly exists?

The answer to this question in complicated. If we decide to only consider expressions of normative bioethics that are highly valued in the west, particularly academic books and articles, we may, indeed have to give an affirmative answer to this question. However, by doing so we run the risk of overlooking the possibility that Hindus develop and practice normative bioethics without writing it down and publishing it. Earlier empirical research among healthcare professionals, although not entirely unequivocal in its conclusions, points to this possibility by showing that Hinduism influences attitudes to ethical issues in healthcare. A study conducted in 2010 among physicians practicing in four teaching hospitals in the South-Indian city of Mangalore showed that Hindu doctors were less likely ($p = 0.001$) to approve euthanasia than their Muslim and Christian colleagues: among the Hindus, only 39.2% considered euthanasia ethically justified, while 69.4% of Muslims and 67.9% of Christians approved euthanasia (Subba et al. 2016). Another study came to somewhat different observations. The researchers involved in that study assessed the euthanasia attitudes of physicians working in a South-Indian tertiary hospital and compared the attitudes of the respondents with their religious affiliation. The researchers observed a significant association ($p < 0.001$) between religious affiliation and disapproval of euthanasia. Christians and Muslims were much more likely to be against euthanasia. 76.9% of Muslims and 64.3% of Christians were against euthanasia, while only 24.3% of Hindu respondents disapproved of the practice (Kamath et al. 2011).

On the basis of these studies, it is impossible to reach any definite conclusions regarding the influence of Hinduism on attitudes to euthanasia. This is not only because the final conclusions contradict each other, but also because we do not know how the respondents interpreted euthanasia. In both studies, the term euthanasia was used in a broad undefined sense, which can include forgoing treatment, the administration of a lethal injection, but also intensive pain management that could possibly hasten death. The different outcome could very well be a consequence of differences in interpretation of euthanasia by the study participants. What these studies do show, however, is that Hinduism may, indeed, influence the ethical attitudes of Hindus. If this is the case, we should attempt to determine how Hinduism directs the bioethical attitudes of Hindus and to what extent it enables them to develop normative bioethical arguments. In the remainder of this chapter, I will formulate an answer to these two questions.

To this end, I will rely on empirical data. By doing so, my approach substantially differs from that of earlier academic studies on Hindu bioethics, such as the two works by S. Cromwell Crawford, to which I already referred above: *Dilemmas of Life and Death: Hindu Ethics in North American Context* (1995), and *Hindu*

Bioethics for the Twenty-First Century (2003), and Katherine K. Young's article *Euthanasia: Traditional Hindu Views and the Contemporary Debate* (1989), that I also briefly mentioned above. To be precise, these are not the only academic works on Hindu bioethics: there are many brief discussions and introductions to Hindu bioethics mainly in the shape of journal articles or articles in collected volumes. Yet, the works by Crawford and Young are definitely among the most comprehensive, and, clearly, the most influential studies on Hindu bioethics, as is illustrated by the number of times they have been cited in other publications on Hindu bioethics. In these works, the authors analyze traditional Hindu texts, philosophical treatises, and the views of a few notable Hindu thinkers, but include no data directly collected among contemporary Hindus. In this chapter, I will include observations from three empirical studies that were conducted in India and involved Hindu samples; that is people who self-identified as Hindus. These empirical studies provide us with data that will help us to understand the nature of normativity in Hindu bioethics. The empirical data were all collected in different palliative care programs in New Delhi over the past 10 years.

The first empirical study that will be included in the current analysis was conducted from February to November 2008 among 27 physicians and nurses working in palliative care programs in New Delhi. It was an interview study based on Grounded Theory methodology. It assessed the influence of religion on the participants' attitudes to treatment decisions at the end of life (Gielen et al. 2011a; Gielen et al. 2011b). The second empirical study took place over the course of 2012 at a pain and palliative care unit in a tertiary hospital in New Delhi. It was an ethnographic study that more broadly analyzed the role of religion and spirituality in that palliative care context (Gielen 2014; Gielen and Kashyap 2019). The third empirical study consisted of a questionnaire-based survey of 300 patients attending a pain clinic of a tertiary hospital in New Delhi. The questionnaire focused on spiritual issues and concerns (Bhatnagar et al. 2016; Gielen et al. 2017). The methodologies of these studies have been described elsewhere and, for the purpose of the current theoretical study, it is neither necessary nor helpful to repeat them here.

In what follows, I will analyze the empirical data in order to describe the nature of normative argument development regarding bioethical issues in Hinduism. In this way, I will be able to show that normative reasoning in Hindu bioethics combines two ethical theories: virtue ethics and deontological ethics. In the next section, I will analyze aspects of Hindu bioethics that point to virtue ethics as a relevant ethical theory to understand Hindu bioethical reflection. In the subsequent section, I will present tendencies in Hindu bioethics that seem to be more akin to deontological ethics. In the last section before the conclusion, I will argue that the combination of these two ethical theories is assumed in the philosophical concept *saṃskāra*. Throughout this exploration of normativity in Hindu bioethics, I will focus on the particular issue of euthanasia to illustrate how normative ethical arguments are developed in Hinduism. Euthanasia is defined as "the intentional administration of lethal drugs in order to painlessly terminate the life of a patient suffering from an incurable condition deemed unbearable." (Broeckaert 2009) Euthanasia is a prominent issue in healthcare ethics and, as such, it is a relevant example for our analysis of normative ethical argument development in Hinduism. Relevant insights regarding euthanasia attitudes in Hinduism can also be derived from the empirical data that were collected in New Delhi. In that regard, euthanasia is a meaningful issue to compare observations from the academic literature and the empirical data.

6.3 Hindu Bioethics as Virtue Ethics

The term 'virtue ethics' covers a group of ethical theories that are not primarily interested in whether or not persons act in line with duties that confirm to norms, as in deontological ethics, or whether or not their acts lead to good consequences, as in consequentialism. Virtue ethics rather emphasizes the development of virtuous character traits. In this way, virtue ethics does not give a direct answer as to whether specific acts are right or wrong. In virtue ethics, moral judgement implies an assessment of the motives that underlie actions. In a virtuous person, motives express virtues. Thus, virtues are character traits that motivate persons to do good actions. Character traits differ from person to person and gradually develop over time. Likewise, a virtuous character is the outcome of moral growth which takes place starting from innate dispositions through human interaction, experience, and education.

As an ethical theory, virtue ethics is the product of Western philosophical thinking and its roots can be traced back to Aristotle. When we survey Hindu thinking and culture for normative ethical strands, we should not expect to find historical connections or even exact parallels with virtue ethics. After all, until Western colonial powers substantially expanded their power in India from the seventeenth century onwards, Indian thought had been developing largely independently of Western philosophy. This means that similarities with virtue ethics are, in a sense, coincidental. At the same time, these similarities are not a mirage and a critical assessment of the extent to which Hindu bioethics resembles virtue ethics may help us to understand the former's unique character, and, from there, the construction of normativity within it.

The Hindu belief in karma is probably the best example to illustrate how Hindu reflection on bioethical issues displays tendencies that resemble virtue ethics. Above, I already explained that belief in karma refers to the view that every living being will experience the consequences of his or her actions. In Hinduism, it is generally accepted that consequences of actions will be experienced in the present life or in a future life. In this way, karma is related to the belief in rebirth, called *punar-janma* in Sanskrit. Good deeds will lead to good consequences, and, thus, eventually, a good rebirth, while bad or evil deeds guarantee bad consequences, and, sooner or later, a bad rebirth.

By demanding good actions as a prerequisite for good consequences the karma belief could stimulate moral action. The question, then, is which actions are good, and which ones are bad? To answer this question, we have to determine whether Hinduism provides a clear and unambiguous account of which actions will lead to good karma or, to say it differently, which actions are moral. If Hinduism does, what kind of answers does it provide? Crawford, one of the above-mentioned scholars of Hindu bioethics, is of the opinion that a careful analysis of the Hindu tradition enables a scholar to reach definite conclusions regarding the permissibility of actions through the application of karmic theory. Crawford shows, for instance, how an assessment of the karma belief leads to the conclusion that euthanasia is morally permissible in Hinduism. In his balanced argument, Crawford considers the diversity of Hindu views on euthanasia. He starts by describing Hindu views that state that euthanasia is not permissible because it cuts short a life that, through suffering, would have offered an opportunity to atone for bad karma that had been accumulated throughout this life or previous lives. Crawford ultimately rejects such views arguing that they lead to unjustified fatalism and Hinduism also believes in the possibility of offsetting bad effects through good or right deeds. From there, Crawford concludes that a choice for euthanasia can be a right action in the case of extreme physical suffering at the end of life, just as taking medicine that can cure a disease is a right action, too (Crawford 1995, pp. 109–114).

Such a conclusion can, of course, difficultly be called virtue ethics, because Crawford focuses on whether or not the action in itself is right or wrong. To be true, in Crawford's argument, a person's character, and related aspects such as intentions and motives are not entirely out of the picture, but they are definitely not at the heart of the analysis. The kind of conclusions that he reaches are more in line with deontological ethics, which we will discuss in the next section. We have to wonder, though, whether the concept of karma as an ethical category is suited to reach the kind of conclusions that Crawford obtains through it. Reflections by a very influential figure like Mahātmā Gandhi suggest that we cannot be very confident that it actually does.

On November 28, 1928, Gandhi responded in a letter to a man who had written to him about his 20-year old sister. She had been suffering from paralytic attacks. Various treatments had failed to be effective. She had been considering suicide and had asked her brother whether he would allow her to kill herself. Finally, she decided against committing suicide because her suffering would offer her an opportunity to purify her karma. Because this young lady had been planning to kill herself, the issue at hand is not euthanasia. If a physician would prescribe lethal drugs which she would subsequently administer herself, we would be talking about assisted suicide. The case is, nevertheless, relevant because the argument that she presented is similar to the one that Crawford rejected. So, Gandhi's answer is relevant. Initially, Gandhi's response seems to go into the same direction as Crawford's. Gandhi reprimanded the woman for considering karma a mechanic principle. According to him "the law [of karma] is not a dead, rigid, inert thing, but it is an ever-living, ever growing mighty force." With this answer, Gandhi indicates, just like Crawford, that people should take their destiny into their own hands and should not fatalistically accept suffering as a consequence of bad karma that has to be endured, but, unlike Crawford, Gandhi did not conclude from there that actively seeking death could be warranted.

Gandhi suspected that the young woman's desire to commit suicide was based on the fear that she would be a nuisance to those who had to care for her daily. He explained that she should not have this kind of feelings, because it was simply the duty of those around her to care for her. He concluded: "What she considered was an inconvenience to her nurses was a privilege, or should be, in the latter's estimation. And if she desired death, the nurses could not comply with her desire, for that compliance would be tantamount to shirking of an obvious duty." (Gandhi 1958–1994, vol. 38, pp. 108–109) By saying this, Gandhi means that the lady's relatives should care for her and, more generally, display a caring attitude. In his view, the suffering young woman should not feel guilty about the care that she receives, because it is her relatives' duty to care for her. Gandhi does not specify why it is their duty, or why that duty would trump the woman's autonomous decision or desire for death. He simply assumes that caring is what good or, we could also say, virtuous people do. Gandhi emphasizes good character and any consideration of eventual karmic consequences seems to be irrelevant in his argument. So, here, 'duty' is not to be understood in a deontological sense of specific norms that have to be followed in actions, but rather as a call to live virtuously.

In Indian thought, the connection between duty and virtue is not at all unique to Gandhi. The Sanskrit term for duty is *dharma. Dharma* is a central concept in Hinduism. For a Hindu, it is essential to live according to his or her *dharma* to obtain redemption (*mukti* or *mokṣa*) from the cycle of death and rebirth, because acts that are against *dharma* will lead to bad karma, which will further remove the individual from reaching the ultimate goal of redemption. Yet, the concrete content of *dharma* tends to be less clear. Partially, the lack of clarity regarding the concrete content of duties in Hinduism has to be explained by variations and contradictions between the texts that stipulate these duties within the *dharmaśāstra,* which literally means the "science of *dharma*" and refers to the corpus of Indian law. More generally, however, the problem is that, in many concrete cases, Hindu texts do not describe what an individual's duty is. This is particularly true for modern dilemmas such as euthanasia. Whenever this is the case, how should people know their duty?

In the Manu Smṛti (Manu II, 6 & 12), a well-known text that belongs to the *dharmaśāstra,* we find the following answer to the question on the sources for knowledge of dharma:

> The root of *dharma* [consists of] the whole Veda, and the *smṛti* [or traditions] and customs of those understanding these things, and indeed the conduct and the very satisfaction of the *ātman* of the righteous ones. They have declared this fourfold [source existing of] the Veda, the *smṛti,* the good custom, and the pleasure of the own *ātman* an indication of the personified incarnation of *dharma.*

In this quote, Hindus are admonished to first look for what their duties are in religious texts, particularly the Veda, but also other traditional texts. Being aware that, often, relevant answers will not be found in these texts, the authors add that people should act according to the "customs of those understanding these [religious or ethical] things" and the "pleasure of the own *ātman* [or soul]." The implication of this statement is that Hindus should follow the example of virtuous persons and act along the lines of the values or virtues that are ingrained in their soul. Such a recommendation, indeed, closely resembles virtue ethics.

This is, in a sense, also what Clooney observed when he studied the development of bioethical arguments in Śrīvaiṣṇavism, a branch of Hinduism that is primarily found in South India. Clooney noted that this tradition has little to offer in terms of direct answers to bioethical issues. What was emphasized by writers in this tradition, though, was "character formation." Believers had to master the virtues that are embedded in their sacred texts. These virtues would enable them to act ethically in the outside world and respond properly to complicated bioethical issues (Clooney 1995). The question that arises, then, is which virtues should guide behavior.

A survey of Hindu texts would reveal several virtues that could be relevant to bioethics. One such virtue that is of particular relevance to bioethics and healthcare is 'care'. Care is an important virtue in Indian culture and society. Indian society expects people to care not just for their children, but also for their parents, and even, if the need arises, for more distant relatives. This virtue of care is so central in Indian culture that in the first half of the twentieth century Hindu reformers took recourse to it to reinterpret the phenomenon of caste, which has religious connotations, as well. In their view, caste divided society in groups that needed and depended upon each other because each of these groups had specific functions that were essential to society and could not be performed by members of other groups. The performance of these functions was the way in which individuals cared for each other. Society could only thrive due to this care (Gielen 2013, pp. 85–89).

The centrality of the ideal of care in Indian culture and society is the reason why it can be claimed that care ethics is a suitable ethical theory for the Indian healthcare context (Gielen 2016, pp. 20–23). Care ethics, also called ethics of care, is an ethical theory that has been argued by some to be a kind of virtue ethics. Care ethicists contend that everyone, even the most independently appearing person, has caring needs, for the fulfillment of which they have to rely on others. By needing care and fulfilling care needs of others all human beings are interconnected. According to care ethicists, care is the basis of human connectedness and is essential to human lives.

One need not necessarily agree with the suitability of care ethics as a relevant ethical theory for the study and development of bioethics in India in order to, nevertheless, grasp the importance of the virtue of care in that context. I also encountered this emphasis on care when I discussed the issue of euthanasia with palliative care Hindu physicians and nurses. When they explained why they approved or disapproved it, they did not speak about the effect of euthanasia on the patient's karma, but they analyzed whether or not the administration of a lethal injection to a terminally ill patient who had requested it would lead to bad karma for the physician or nurse who was going to administer it. A few did believe that this act would result in bad karma for the physician or nurse, because, in their view, it equaled murder. However, others argued that it would not lead to bad karma, because the lethal injection was administered for the benefit of the patient and his or her relatives. One doctor said:

> That is not bad karma ... because we're relieving someone's pain, while knowing that he has only a little bit of time left and the pain will not go away and the patient also wants it. There is nothing wrong with that.

Another physician argued:

> You're administering euthanasia for the benefit of the patient and his relatives. That's why I don't think that the result for you will be bad.

From these statements we can derive that, in the view of these participants, euthanasia is not moral or immoral by itself. The morality of the act is determined by the motivation that underlies it. We need to assess to what extent the act is a genuine expression of care. If the intention is to care for the patient by reducing his or her pain, than euthanasia could be permissible. At the same time, physicians and nurses emphasized that proper care would almost always lead to adequate symptom control. They argued that the medical team should not only pay attention to physical pain, but also to the patients' psycho-social and spiritual suffering, like loneliness and the feeling that God has forsaken them. Therefore, it is critical that the team assures the patients that they are there to support them. One nurse said:

> We have to be with them, and we have to listen to them. Then with empathy and compassion we have to tell them, we have to explain them: "See, we are with you. We will help you by giving comfort."

This statement obviously makes us to recall Gandhi's advice to the suffering young women when he advised that she should let her relatives care for her. What truly mattered to Gandhi was the act of caring. In this context, arguments related to the belief in karma were not of great importance to him because the outcome of a reasoning based on that belief was unclear. Likewise, among the physicians and nurses, we see that the idea of karma can be used to argue against euthanasia, but also to defend it. Therefore, we should not be too surprised that they were unsure or denied that the idea of karma played a significant role when they took concrete ethical decisions in medical practice. One of the physicians who had argued that the administration of a lethal injection would lead to bad karma for the involved physician or nurse, added that reflections about karma will not always play a role when decisions are taken.

> All of the time this thought of karma doesn't come. ... It may come, but not all the time.

This rather limited role of belief in karma in concrete ethical decision making confirms that this belief functions in the first place as an abstract stimulator of virtuous behavior, and not so much as a guide for concrete action. Although the idea of karma did not always play a role, as this quote shows, the virtue of care was prominent. However, as we also saw, there was disagreement about whether or not virtuous caring could include euthanasia. Given this ambiguity, we should not be too surprised that in certain situations Hindus search for a different kind of guidance, which they may find in modes of deontological ethics.

6.4 Hindu Bioethics as Deontological Ethics

Deontology focuses on norms. In deontological ethics, actions that conform to these norms are right or moral. In Western culture, the biblical command "Thou shall not kill" is the most well-known expression of a deontological norm. The statement professes that killing is wrong, so people should act in such a way that they do not kill. In this way, deontological ethics may seem very straightforward and lead to upfront conclusions. To be true, deontological arguments will not always be simple. For instance, there may be certain situations in which actions that lead to the death of a person may nevertheless be morally permissible, such as cases in which it is necessary to kill one person in order to save two or more others. Through its avoidance of simplistic conclusions deontological ethics tries to incorporate the complexities of human life, just as virtue ethics. The difference between deontological ethics and virtue ethics is that the former directly tells us what we ought to do, while the latter assesses what kind of person we ought to be through the virtues that we express in our actions.

In this vein, we saw that, in Hindu virtue ethics, the proper answer to a euthanasia request is care. This approach does not inform us whether euthanasia is moral or immoral, because euthanasia may or may not be an expression of a caring attitude depending on the particular context and the motives of the person administering euthanasia. We already pointed out that Crawford attempted to construct a deontological Hindu argument by relying on the Hindu belief in karma. The argument that he developed was not very convincing because karma makes more sense if it is interpreted within Hindu virtue ethics. However, this should not be taken to mean that in Hindu ethics there never is a connection between karmic thinking and deontological reasoning. This can be seen in the above mentioned arguments of physicians and nurses who said that euthanasia will lead to bad karma for the physician who administered it. Their reasoning was that euthanasia equals killing, which is impermissible or immoral and will, therefore, lead to bad karma. The condemnation of killing was deontological in nature. The physicians and nurses were able to make a mental connection between deontological assessment of euthanasia and karma, but the deontological conclusion about the immorality of killing was not based on a careful consideration of karmic theory in the line of Crawford. In fact, it was reached without taking recourse to the concept of karma at all. Belief in karma did not motivate the conclusions.

When the physicians and nurses expressed disapproval of euthanasia they frequently mentioned the belief that human life has been given and, therefore, humans do not have the right to take away human life. The interviewed physicians and nurses put euthanasia on par with ending a human life prematurely. Since they saw human life as a gift from a power or force transcending human existence, taking away a human life before its time is unacceptable. A physician and a nurse, who were asked about their view on euthanasia, did not further clarify which transcending force or power has given life to humans. They just took for fact that human beings do not create life on their own, and, therefore, have no right to end it prematurely. One physician said:

> I think, once we have got life we should live the way we have got it. We shouldn't try to interfere, shorten, or increase it like that if we can't give life to anybody.

A nurse said:

> According to me, we cannot give life. We should not be the ones who kill. … If we cannot bring someone to life, then we also should not kill anyone. For that reason, I'm not in favor of euthanasia.

The remaining nurses and physicians who argued against euthanasia on the basis of the conviction that life has been given to humans were clearer regarding the nature of the force or power that grants human life. One nurse still doubted whether human life is given by God or nature.

> If we cannot give him [the patient] life, then we also shouldn't kill him. We can relieve his pain. We can give him more medicine. But we shouldn't kill him. … Human life is in God's hand. That right is not in our hands. It's in nature, God, whatever it is.

Other interviewees firmly stated that human life has been given by God. Since human life has been given by God, the right to take away human life uniquely rests with God. This is what three of them said:

> God gives life. Because of that, God can also take it. We cannot take life.
> No one except God has the right to take away one's life.
> Human beings do not have the right to take life that has been given by God.

In this way, almost all interviewees who reported to believe that life has been given by a higher Power connected this conviction with a belief in a God who is the only entity having the right to decide about life and death. For the physicians and nurses, this belief was the central element in their deontological argument against the permissibility of euthanasia.

The belief in a personal God who has the ability to actively steer and determine human existence is a prominent aspect of contemporary Hinduism and it is, therefore, not strange at all to observe its significance in Hindu ethical reasoning. Hindu gods and goddesses are believed to be all-powerful. Hindus also believe that these goddesses and gods will use their powers to protect their faithful devotees. Therefore, they are often portrayed with their right hand upright, and the palm turned outwards. This gesture represents the *abhayamudrā:* the sign of removal of fear. God is seen as almighty and even decides about life and death. Yet, for those who have focused their life on Him/Her fearing either God or any adversity is unnecessary because God will protect them against evil.

In healthcare, such a belief can be very powerful. Above, the quotes of the physicians and nurses illustrated how it directed the ethical attitudes of physicians and nurses to euthanasia. The belief has also a strong impact on how Indian patients experience their disease. In the ethnographic study at the pain and palliative care unit, many patients were found to derive hope for a cure from their strong faith in God. They saw themselves as people who throughout their life had been faithful to

God. They held the belief that God would finally listen to their prayers and eventually grant cure. The truth was that several of these patients were terminally ill and nearing death. These patients felt at a loss when they finally realized that the hoped for cure would not materialise. For some of them, this spiritual crisis contributed to a depression that had been lingering. Several of these patients directly or indirectly indicated that they wanted to die, rather than living with their disease. "If God does not want to cure me," one patient said in despair, "He can take me up." The spiritual disillusion caused by the feeling that God had left them when they needed Him most, combined with the inability to find a satisfactory answer to existential questions, such as "Why am I not getting better?" and "Why does God not cure me?" made patients to be angry at God, or even wonder whether He exists at all. Some patients testified, for instance, that they were no longer able to pray because of anger or frustration with God (Gielen 2014; Gielen and Kashyap 2019).

The importance of the belief in an all-powerful God who cares for his devotees and has the ability to cure them was also confirmed in the quantitative study among patients attending a cancer pain clinic in New Delhi. The survey findings showed the desire of the patients to experience divine support. 88.3% agreed with the statement "God is with me." 87% testified to have a belief in God that gave them strength. 85% was of the opinion that their faith supported them in their illness and 78.9% expressed their belief in a higher power that supported them. Likewise, a large majority of the patients was interested in activities that helped them to connect with God, such as prayer and devotional acts. 90.3% said that praying or chanting made them feel better and 83.3% felt that *pūjā,* a ritual expression of devotion to God, was a source of peace to them (Gielen et al. 2017).

These data were not yet available when Crawford wrote his seminal works on Hindu bioethics. So, he can easily be forgiven for not paying more attention to the belief in a powerful personal God in his writings on Hindu bioethics. On the other hand, our data suggest that this belief is a guiding principle for a majority of Hindus in healthcare in India, both patients and healthcare providers. It impacts experience of illness and ethical attitudes. Therefore, this belief in a personal God should not be overlooked in a critical assessment of Hindu bioethics.

For our current analysis, the idea is of particular relevance because it supports a clearly deontological argument on euthanasia and shows that Hindu bioethics has outspoken deontological tendencies, too. In this sense, the presented data are an empirical verification of Crawford's assumption that systematic deontological arguments are a possible outcome of Hindu bioethics. Crawford showed what such arguments might look like. Since Crawford did not verify to what extent his theoretical model corresponded to ethical reasoning by Hindus in real life, he overlooked the possible deontological implications of belief in a personal God among Hindus and overestimated the capacity of belief in karma to direct deontological conclusions. Moreover, his interpretation of karma as a source of deontological reasoning seems to have led him to conclusions that differ from those normally reached by Hindus. Crawford's deontological reading of karma illustrates his limited interest in the virtue-ethical aspects of Hindu bioethics. The outcome of his approach was a representation of normative Hindu bioethics that was somewhat biased towards deontological ethics and underestimated the import of virtue ethics.

Although Crawford's representation misses out on virtue ethics as an essential aspect of Hindu bioethics it has to be admitted that it was not all unreasonable from Crawford to take a deontological approach to Hindu bioethics because, in the Hindu tradition, the separation between deontological ethics and virtue ethics is not made and it may, therefore, be tempting to emphasize only one of them in an analysis of Hindu bioethics. This is illustrated by the way in which Hindus interpret the concept *saṃskāra*. As a philosophical idea, this notion has ethical implications that should be considered while assessing Hindu bioethics.

6.5 Hindu Bioethics as Based on *Saṃskāras*

In Hindu philosophy, *saṃskāra* refers to the dispositions or character traits that make living beings do particular actions. *Saṃskāra*s are not necessarily static. By regularly doing certain actions or not doing certain actions *saṃskāra*s are conditioned and can, then, more profoundly guide behavior. Interestingly, *saṃskāra*s also function at an epistemological level and have the power to impact religious knowledge. Hindu philosophers have reflected upon the ways in which this happens (Dalal 2012; Mishra 1953; Whicher 2005). However, if we want to understand how the concept relates to ethical normative arguments development by Hindus in biomedicine, we have to look at what that idea means to Hindus who develop such arguments. What we have observed earlier is that these Hindus are generally not persons who write elaborate ethical treatises, but rather individuals who, in their day-to-day life, are confronted with the ethical issues that arise within the context of modern biomedicine. Examples of such persons include the palliative care physicians and nurses among whom I collected data.

They did speak about *saṃskāra*s and defined these in three ways: habits (1), good or bad works (2), and effect of one's upbringing (3). A few Hindu nurses and physicians described *saṃskāra*s as habits. These habits could be good or bad, as one physician clarified.

> *Saṃskāra* means by repeatedly doing a particular action, we will in some way get used to this particular action. And then that will become a sort of habit. And that is what we call *saṃskāra*, good or bad.

In line with this, another physician more generally interpreted *saṃskāra*s as good or bad works.

> *Saṃskāra* we call any good or bad thing that has been done by you. If you're doing a good work, then we say that that is a good *saṃskāra*. You have done a good *saṃskāra*. And if you do a bad work, then we say this is a bad *saṃskāra*.

Most physicians and nurses interpreted *saṃskāra*s as things people learn through their upbringing in a family. In that sense, *saṃskāra* means the effect of one's upbringing. In the words of three different nurses:

> *Saṃskāra* means how you have been brought up in the family, and what you have been taught from your childhood onwards.

For us, *saṃskāra* means whatever we have been taught at home from our childhood onwards, from the beginning onwards.

Saṃskāra is whatever I have learned in my family and from my background.

This effect of upbringing was primarily seen as something positive. Through education, people learn which values are important in life. In this way, *saṃskāra*s were more generally interpreted as values that enable people to distinguish between good and bad. Two physicians further argued that these values show people which direction they should go in life and how they should or should not behave.

Saṃskāra means teaching that has been given to you about life, what is good, what is bad, what you have to do, what you should not do.

Saṃskāra means doing good things, not dealing with bad things, and doing your work well, sincerity.

A physician added that upbringing enables people to interiorize these values, so that they do not harm others.

*Saṃskāra*s make you learn good things. Good deeds, good work, good feelings. […] If you have got good *saṃskāra*s this means that our parents have taught us: "Child, don't do anything that may harm another person, or that may cause loss to someone. Or, don't lie." Then this *saṃskāra* becomes part of us. Then we come to realize that if we lie, we're doing something wrong, or if I harm someone, it will hurt me somewhere.

These responses interest us here in particular because they show that these physicians' and nurses' interpretation of *saṃskāra* includes aspects of both virtue ethics and deontological ethics. A physician who was quoted above spoke about "sincerity," which can be seen as a virtue. Other healthcare providers referred to the principle of non-maleficence, or do-no-harm, and one mentioned "don't lie." These are deontological statements.

The coexistence of virtue ethics and deontological ethics in these interpretations of the concept *saṃskāra* underscores the particular nature of Hindu bioethics. The particularity of normative reasoning in Hindu bioethics does not just reside in the inclusion of Hindu concepts and ideas, such as karma, in ethical arguments. It equally lies in its application of a not consciously articulated ethical theory that encompasses virtue ethics as well as deontological ethics. For Hindus who are developing bioethical arguments, virtues of virtue ethics and norms of deontological ethics are all, in a sense, relevant values that have to be considered while attempting to formulate answers to the ethical challenges that biomedicine poses. The advantage of this approach is that it can stimulate Hindus to look at bioethical issues from different perspectives. The disadvantage is obviously that it is not clear which conclusion should receive priority in case these perspectives lead to conflicting findings. In our research, for instance, we observed that a majority of Hindu physicians and nurses disapproved euthanasia. But, that does not mean that we can disqualify as unauthentic as per Hinduism the views of those Hindus who approved of the practice. The fact that Hindus have differing values and *saṃskāra*s will inevitably lead to conflicting conclusions regarding bioethical issues.

Normativity in Hindu bioethics does, therefore, not reside in particular ethical judgements, but in the search for ethical answers itself. This means that an ethical conclusion has normative value from a Hindu perspective if it is reached by following the Hindu approach to bioethics, which implies the application of Hindu religious concepts and ideas to the bioethical problem at hand through the combined lenses of virtue ethics and deontological ethics.

Such a flexible approach to normativity should enable Hindus to be open to a wide range of arguments when they are confronted with complex ethical issues in healthcare and biomedicine. They may still have a strong and unwavering opinion on certain issues. Above, we saw that many Hindus are firmly convinced that euthanasia is ethically unacceptable because the person who administers the lethal injection prematurely ends a life that has been given by God. At the same time, that conviction will most often not be of a fundamentalist nature, which would preclude Hindus from considering alternative points of view. Even more, in line with their tradition, Hindus should be willing to respectfully consider arguments that include aspects of religions other than Hinduism. Again, finally, they need not agree with these arguments. Yet, given the fact that, in Hinduism, we find the coexistence of beliefs and practices that are extremely diverging and not rarely contradictory, Hindus would be inconsequential if they did not extend their willingness to consider diverging religious points of view beyond the borders of their own religion.

In this way, we can ask ourselves whether Hinduism does not provide a very relevant approach that could inform a global view on normativity in bioethics. Hinduism shows us that normativity should not primarily be focused on the outcome of the ethical argument, but rather on the process towards that outcome. In contemporary multicultural society, normativity becomes a problem if we emphasize the outcome. The outcome of an ethical argument is obviously very important because it determines how people will act. However, if the emphasis on the outcome leads to the conclusion that only one answer can be normative, the result will be that conclusions that diverge from the own answer will be seen as either bereft of normative value, or as normative for the individual who developed that view within the framework of a specific ideology. The former is a fundamentalist ethical absolutism, while the latter is ethical relativism. The Hindu approach that focuses on the ethical process rather than the outcome shows us that it is possible to develop a bioethical argument towards one outcome and defend that outcome as the preferred position while at the same time recognizing the normative value of diverging outcomes. The condition is that these outcomes have been arrived at through a critical analytical process that involves the consideration of various perspectives.

6.6 Conclusion

The above analysis has shown that Hindu normative bioethics entails more than applying Hindu concepts and norms to bioethical issues. Hindu beliefs and ideas are, obviously, an essential part of normative ethical reasoning in Hinduism, but these religious concepts on their own do not give an argument normative value from a Hindu perspective. Normativity in Hindu bioethics implies a particular method that includes a deontological approach and virtue ethics. The fact that Hindu bioethics includes the application of two different ethical theories in its search for normative answers to bioethical issues indicates its methodological openness to the consideration of diverging perspectives. This makes Hindu bioethics into an interesting example of how normativity could be an applicable concept in global bioethics were diverging cultural and religious backgrounds seem to make normativity a highly problematic notion. Hinduism teaches us that for the determination of normativity in bioethics the process towards the conclusion is more important than the conclusion itself. Arguments can be considered valid or normative when they have approached the issue from various theoretical perspectives. Such approach may indeed contribute to a better appreciation of contending arguments in bioethics without making preference for a particular argument impossible.

Suggested Readings

The most well-known and influential works on Hindu bioethics are two books by S. Cromwell Crawford and a book written by Harold Coward, Julius J. Lipner, and Katherine K. Young.

Crawford, S.C. 1995. *Dilemmas of Life and Death. Hindu Ethics in a North American Context*. New York: State University of New York Press.

Crawford, S.C. 2003. *Hindu Bioethics for the Twenty-first Century*. Albany: State University of New York Press.

Coward, H., J. Lippner, and K. Young. 2000. *Hindu Ethics. Purity, Abortion and Euthanasia*. Albany: State University of New York Press.

A critical analysis of the approach used by the authors of these works can be found in the following article by Francis X. Clooney.

Clooney, F.X. 1995. Back to the Basics: Reflections on Moral Discourse in a Contemporary Hindu Community. *The Journal of Medicine and Philosophy* 20 (4): 439–457.

References

Balagangadhara, S.N. 1994. The Heathen in His Blindness…. In *Asia, the West and the Dynamic of Religion*. Leiden/New York/Keulen: Brill.

Bhatnagar, S., S. Noble, S.K. Chaturvedi, and J. Gielen. 2016. Development and Psychometric Assessment of a Spirituality Questionnaire for Indian Palliative Care Patients. *Indian Journal of Palliative Care* 22 (1): 9–18.

Broeckaert, B. 2009. Euthanasia and Physician Assisted Suicide. In *Palliative Medicine*, ed. D. Walsh, 110–115. Philadelphia: Elsevier.

Clooney, F.X. 1995. Back to the Basics: Reflections on Moral Discourse in a Contemporary Hindu Community. *The Journal of Medicine and Philosophy* 20 (4): 439–457.

Coward, H., and T. Sidhu. 2000. Bioethics for Clinicians: 19. Hinduism and Sikhism. *CMAJ* 163 (9): 1167–1170.

Crawford, S.C. 1995. Dilemmas of Life and Death. In *Hindu Ethics in a North American Context*. New York: State University of New York Press.
———. 2003. *Hindu Bioethics for the Twenty-first Century*. Albany: State University of New York Press.
Dalal, N. 2012. Clouding Self-Identity: Sankara, Samskaras, and the Possession of King Amaruka. *The Journal of Hindu Studies* 5 (3): 283–292.
Deshpande, O., M.C. Reid, and A.S. Rao. 2005. Attitudes of Asian-Indian Hindus Toward End-of-life Care. *Journal of the American Geriatrics Society* 53 (1): 131–135.
Gandhi, M. (1958–1994). *The Collected Works of Mahatma Gandhi*. Delhi: Government of India, Ministry of Information and Broadcasting, Publications Division.
Gielen, J. 2013. *God in India: diversiteit en devotie in de Indische religies*. Tielt: LannooCampus.
———. 2014. Why Does Spirituality Matter in End of Life Care? In *Concepts of Palliative Care*, ed. R. Das and S. Nayak, 72–80. Cuttack: A.K. Mishra Publishers.
———. 2016. Education in Care Ethics: A Way to Increase Palliative Care Awareness in India. *International Journal of Ethics Education* 1 (1): 15–24.
Gielen, J., S. Bhatnagar, S. Mishra, A.K. Chaturvedi, H. Gupta, A. Rajvanshi, et al. 2011a. Can Curative or Life-Sustaining Treatment be Withheld or Withdrawn? The Opinions and Views of Indian Palliative-Care Nurses and Physicians. *Medicine, Health Care, and Philosophy* 14 (1): 5–18.
Gielen, J., H. Gupta, A. Rajvanshi, S. Bhatnagar, S. Mishra, A.K. Chaturvedi, et al. 2011b. The Attitudes of Indian Palliative-care Nurses and Physicians to Pain Control and Palliative Sedation. *Indian Journal of Palliative Care* 17 (1): 33–41.
Gielen, J., S. Bhatnagar, and S.K. Chaturvedi. 2017. Prevalence and Nature of Spiritual Distress Among Palliative Care Patients in India. *Journal of Religion and Health* 56 (2): 530–544.
Gielen, J., and K. Kashyap. 2019. Belief in Karma and Mokṣa at the End of Life in India. In *Death Across Cultures: Death and Dying in Non-Western Cultures*, ed. H. Selin and R.M. Rakoff, 19–39. Cham: Springer.
Gudwani, A., P. Mitra, A. Puri, and M. Vaidya (2012). India Healthcare: Inspiring Possibilities, Challenging Journey. http://www.mckinsey.com/. Retrieved from http://www.mckinsey.com/
Kamath, S., P. Bhate, G. Mathew, S. Sashidharan, and A.B. Daniel. 2011. Attitudes Toward Euthanasia Among Doctors in a Tertiary Care Hospital in South India: A Cross Sectional study. *Indian Journal of Palliative Care* 17 (3): 197–201.
Mishra, N. 1953. Samskaras in Yoga Philosophy and Western Psychology. *Philosophy East and West* 2 (4): 308–316.
Office of the Registrar General & Census Commissioner India. (2016). Census of India Website. Retrieved from http://www.censusindia.gov.in/
Pennington, B.K. 2005. *Was Hinduism Invented? Britons, Indians and Colonial Construction of Religion*. Oxford: Oxford University Press.
Singh, P., R.J. Yadav, and A. Pandey. 2005. Utilization of Indigenous Systems of Medicine & Homoeopathy in India. *The Indian Journal of Medical Research* 122 (2): 137–142.
Staal, F. 1996. *Ritual and Mantras: Rules Without Meaning*. Delhi: Motilal Banarsidass.
Subba, S.H., V. Khullar, Y. Latafat, K. Chawla, A. Nirmal, and T. Chaudhary. 2016. Doctors' Attitude Towards Euthanasia: A Cross-sectional Study Experience. *The Journal of the Association of Physicians of India* 64 (6): 44–47.
UNESCAP-SSWA. (2015). India and the MDGs: Towards a Sustainable Future for All: United Nations.
Von Stietencron, H. 2007. *Hindu Myth, Hindu History: Religion, Art, and Politics*. Permanent Black: Ranikhet.
Whicher, I. 2005. The Liberating Role of Samskāra in Classical Yoga. *Journal of Indian Philosophy* 33 (5–6): 601–630.
Young, K. 1989. Euthanasia: Traditional Hindu Views and the Contemporary Debate. In *Hindu Ethics. Purity, Abortion and Euthanasia*, ed. H. Coward, J. Lippner, and K. Young, 71–130. Albany: State University of New York Press.

Chapter 7
Normativity in Buddhism and Its Application in Bioethics

Soraj Hongladarom

Abstract This paper argues that normativity in Buddhism is constituted by a thoroughgoing kind of naturalism and teleology. An action is good in Buddhism only if it is conducive to the main goal of being a Buddhist, that is, becoming released from sufferings or dissatisfactory conditions (dukkha) thereby gaining a release from the bondage of the cycle of life, death and rebirth (*saṃsāra*). This is a sketch of an ethical theory that I call ethical teleological naturalism. The theory is also pragmatic in nature because the emphasis lies on how to achieve the desired goal. Buddhist ethics is also strongly connected to Buddhist metaphysics and epistemology because in order to gain liberation from the cycle one needs to gain full knowledge of what ultimate reality is like. This interpretation of Buddhist ethical thought is thus akin to that of Damien Keown in one respect, namely that the Buddhist theory is comparable in many ways to ancient Greek ethical theory, but Keown sees Buddhist theory to be similar to virtue ethics whereas I see the theory to be more akin to Stoicism as well as Spinoza's ethical theory. The latter half of the paper discusses how this pragmatic and teleological nature of Buddhist normative theory plays out in bioethics. This is done through a look at a recent debate among two prominent scholars of Buddhist philosophy, Damien Keown and Rupert Gethin, on euthanasia.

Keywords Buddhism · Normativity · Naturalism · Ethics · Euthanasia

7.1 Introduction

Buddhism is a thoroughly goal-oriented religion. Being a Buddhist means that one takes up a practice prescribed by the Buddha, and the goal of the practice is to achieve the final stage of Liberation called *Nirvāṇa*, the state of imperturbability where the practitioner is finally released from the cycle of life and death known as *saṃsāra*.

S. Hongladarom (✉)
Department of Philosophy, Faculty of Arts, Chulalongkorn University, Bangkok, Thailand
e-mail: Soraj.H@chula.ac.th

J. Gielen (ed.), *Dealing with Bioethical Issues in a Globalized World*,
Advancing Global Bioethics 14, https://doi.org/10.1007/978-3-030-30432-4_7

This goal-orientedness has a strong bearing on the main topic of this book. One of the main concerns of the book is to investigate the nature of normativity in bioethics through the eyes of various traditions, including those of the major religions. This chapter contributes to this attempt by showing how Buddhism views the nature of normativity. This serves both the function of providing a perspective on how an ancient tradition such as Buddhism views the nature of normativity, and also how in the Buddhist perspective one can adjudicate among various competing normative claims. I will attempt to delineate how Buddhism answers these two questions. These answers not only aim at describing how Buddhism views the question of normativity, but the Buddhist answer could also contribute to the broader ongoing debate on normativity in bioethics, too. In doing so I discuss a recent debate on Buddhist views on euthanasia by Rupert Gethin and Damien Keown, which is very instructive to understand the construction of normative arguments in Buddhist bioethics.

The answers I will argue for are based on the idea that Buddhism is a thoroughly goal-oriented religion as mentioned earlier. By this, I mean that, in Buddhism, what makes something valuable is not what something is in itself, but whether it can bring about the desired goal as well as how that goal is justified. Since invariably the desired goal is ultimately *Nirvāṇa*, in Buddhism the question of what normativity is can then be simply answered that it is this eventual Goal that defines what is valuable or not. Action is positively valued if it is efficacious in bringing about the desired goal. The simplicity here is deceptive, however. What is difficult to grasp is not only what one has to do to bring oneself to achieve *Nirvāṇa*, but along the path there are many thorny philosophical issues as well, such as how one should conceptualize the nature of reality itself. These latter issues have inspired countless philosophical schools within Buddhism itself. These are highly technical issues requiring getting involved in the specialized vocabulary of the various Buddhist schools and paying close attention to the original texts. As this chapter is not intended for specialist Buddhist scholars, I will not go into those concerns in this paper. My purpose is more focused on one aspect. That is, I will try to show that the Buddhist perspective on normativity is based on how the Buddhist looks at teleology. Buddhist teleology, in short, is the source of an adequate answer to both the question concerning how normativity is justified, and how various competing normative claims can be adjudicated.

Discussion on the nature of Buddhist ethics is a popular topic among scholars in the field. Damien Keown's influential book on the topic[1] opened up the discussion by arguing that there is no theorization of ethics in Buddhism, unlike what happened in the West since the ancient Greeks. What he means is that in the West there have been discussions of the nature of goodness, how to justify good action, and so on. In short, there has been no equivalent of Aristotle's *Nicomachean Ethics* in the entire voluminous corpus of Buddhist literature. Thus, for Keown, philosophers who search for a possible Buddhist ethical theory have to interpret the text and compare it with the standard Western theories, such as deontology or consequentialism.

[1] Damien Keown, *The Nature of Buddhist Ethics* (New York, NY: St. Martin's Press, 1992). See also Georges Dreyfus, "Meditation as Ethical Activity," *Journal of Buddhist Ethics*, available at http://blogs.dickinson.edu/buddhistethics/2010/04/05/meditation-as-ethical-activity/

Instead of ethical theory, what is there in the Buddhist texts are guidelines for the practitioner to follow in order to achieve the result. However, in this paper, I am arguing that by emphasizing the practical nature of the ethical guidelines, Buddhism has a theory of its own, in a way at least. The theory, very basically, says that what is good is constituted by whether an act brings about the desired result (i.e., *Nirvāṇa*) in the end or not, and a large part of how this is possible is explained through a reference to Buddhist metaphysics, of which there is a great deal of material in the literature. I call this proto-theory "ethical teleological naturalism," as we shall see in the following sections, although an adequate treatment of this proto-theory would far exceed the limit of this paper.

7.2 Buddhist Naturalism: How the Goal Is Justified

The teleology in Buddhism is straightforward: an action is good if it brings about desirable results, and is bad if it brings about undesirable results. And, as the desired result is always release from the cycle of life and rebirth, any action that can get one closer to this end result is good. The *Abhidhamma*, one of the "Three Baskets" of Buddhist canonical scripture, divides phenomena into three kinds, namely good (*kuṣala*), bad (*akuṣala*), and indeterminate (*abyākata*) phenomena. The first kind of phenomena are the ones that bring in good results, such as "serenity in sense and thought," "lightness," "plasticity," "mindfulness," and "intelligence," all of which are qualities that are conducive to achieving the results of meditation, which is one of the requirements of attaining the Goal of release from the cycle of death and rebirth. On the contrary, the second kind of phenomena are those that are obstacles to achieving the Goal, and these are the qualities that are opposites to each of the good ones described above. The third kind, the indeterminate ones, produce neither good nor bad results and as such they are not relevant for attaining the Goal. For example, a trivial action such as forgetting something or thinking of something unimportant belongs to the third phenomena. The *Abhidhamma* also states that matter itself belongs to this kind of phenomena, because in itself it is neutral and neither supports nor hinders the attainment. The contrast between the first two kinds of action is discussed in the *Dhammapada*, one of the most famous Buddhist texts, where, at the very beginning, we find the following statement:

> Mind precedes all mental states. Mind is their chief; they are all mind-wrought. If with an impure mind a person speaks or acts suffering follows him like the wheel that follows the foot of the ox.
>
> Mind precedes all mental states. Mind is their chief; they are all mind-wrought. If with a pure mind a person speaks or acts happiness follows him like his never-departing shadow.[2]

---

[2] *The Dhammapada: The Buddha's Path of Wisdom*, transl. Acharya Buddharakkhita (Kandy, Sri Lanka: Buddhist Publication Society, 1995), p. 23. Available at http://www.buddhanet.net/pdf_file/scrndhamma.pdf (retrieved October 31, 2016).

In these sentences, we see that an action done with a pure mind is a good one, one that is certain to bring about good results just as the shadow that never departs from oneself. A good action is one that leads to attainment of the Goal, which is *Nirvāṇa*. This is only possible when the action is done with a pure mind, i.e., the mind that is free from ill will, hatred or egoism. Thus, the pure mind here has the quality of being *kuṣala*, or good state, and the text in the *Dhammapada* is clear in stating that an action done with a pure mind will certainly lead to good results which will be sure to follow.

What this means, of course, is that the Buddhist theory of goodness is always linked to the Goal, which is taken to be desirable for everybody. This is because the Goal is always taken to be the desirable, the best possible end for every sentient being, because achieving the Goal is the only way for the being to avoid suffering completely and permanently, and no sentient being would want suffering upon itself.

A very important key here is that attainment of the Goal, *Nirvāṇa*, is not possible if one does not have a full and complete knowledge of nature. Thus, the Goal is justified by referring to the state of suffering and the fact that no sentient being wants to suffer. 'Suffering' (*dukkha*) is a wide-ranging term in Buddhism, which means not only being in physical pain, but also a state where one is not satisfied with one's own condition, where one desires to get away from the condition in which one is in. In this case, it is almost as if any condition where a sentient being is not satisfied is by definition a *dukkha*; hence it is understandable why no sentient being would want to suffer—it is because any condition that a being wants to escape is defined as suffering in the beginning. Thus, the supreme Goal—to become fully released from all kinds of suffering—would then be a universal goal because it is simply the opposite of something one wants get away from, which means something one wants to strive toward and remain in. In this case, there is no possibility of the Goal being something that a being does not want to attain. Hence, an action that is conducive to achieve the Goal would then be universally good too, where 'good' of course is defined as the quality that is so conducive. It cannot be the case, then, that there be a sentient being that does not ultimately want to attain the Goal, and those beings who apparently do not want to do so do not actually know what their ultimate real interests are.

Now, what is meant by one's own ultimate interests? In Buddhism, this is an objective truth, namely one's own ultimate interest for the sake of which one always strives forward. An objection to this view could be that it presupposes that everyone has the same ultimate interests, which might not be true given the wide variety of choices that are available to people. What is ultimate self-interest for one might not be the same for someone else. However, Buddhism sees this as an illusion. For example, acting selfishly runs counter to the path leading to attainment of the Goal, but there are those who insist that acting selfishly is good for them. However, those latter people do not know their own real self-interest because acting selfishly cannot bring about what they themselves really want deeper down, which is their own lasting happiness.

Buddhist ethics is thus very much connected with Buddhist metaphysics. This makes it similar to ancient ethical systems in the West such as Stoicism[3] and early modern philosophy such as Spinoza's.[4] In Stoicism, the goal is to align oneself with nature and act in accordance with nature. The Stoic sage, much like the Buddhist *arahat* (one who has already attained *Nirvāṇa*), is the supreme paragon of virtue because he is in complete control of passions and lives entirely in accordance with the rational structure of nature. In other words, he has complete knowledge of nature and there is a complete alignment between the knowledge and how he lives. What is good, then, is moral virtue, since it is only virtue that aligns perfectly with my most basic interests. This view is similar to the Buddhist one that we have seen: what is good is what is conducive to the most desirable and the most natural and perfect goal that is available to us. This view is also Spinoza's. The best condition in which a human being could be, according to Spinoza, is a state where one has an intellectual love of God. This is a state where one is also in complete control of one's emotions through a full understanding of how things are in the world. One always guides one's action through reason and not through passions. This is the case in Spinoza's thinking, and also in Stoicism and Buddhism. Thus, this ethical viewpoint is not only an exclusive property of the East, but it has been expressed in Western philosophy as well. It is known as ethical naturalism—what determines whether an action is good or bad is whether the action accords with nature or not, and it is an action that accords with nature that brings about the best life possible for a human being. However, ethical naturalism here is distinguished from modern naturalism, which takes away the role of the best life possible and focuses only on the view that ethical statements can be proven right or wrong. Thus, in order to avoid confusion, we better call the view of Buddhism, Stoicism and Spinozism *ethical teleological naturalism*.

[3] A good introductory source is an article in the Stanford Encyclopedia of Philosophy. See Dirk Baltzly, "Stoicism," *The Stanford Encyclopedia of Philosophy (Spring 2014 Edition)*, ed. Edward N. Zalta, available at http://plato.stanford.edu/archives/spr2014/entries/stoicism/

[4] There are many translations of Spinoza's main work, the *Ethics*. One is by Samuel Shirley; see *Spinoza, Complete Works*, transl. Samuel Shirley and ed. Michael Morgan (Indianapolis, IN: Hackett, 2002); another is by Edwin Curley: see Benedict de Spinoza, *Ethics*, transl. Edwin Curley (Penguin Classics, 1996). My interpretation of Buddhist ethical theory here also differs from that of Keown in this aspect. Though I agree largely with Keown that Buddhist ethics has some affinities with ancient Greek theory, I differ from him in that I believe that Buddhist theory is closer to Stoic ethics rather than Aristotle's kind of virtue ethics. The main difference has to do with moral luck, which seems to be a necessary feature in Aristotle's theory. However, in Stoicism one does not need luck in order to become a sage—a feature that I believe is shared with Buddhism to a certain extent. However, this is a topic that would need to be explored further.

7.3 Comparisons with Other Theories

Ethical teleological naturalism looks similar to another ancient ethical theory, virtue ethics. However, the difference is that virtue ethics focuses on the role of practical wisdom, a kind of practical knowledge that allows one to know what one should do in a given situation. In ethical teleological naturalism (ETN), on the contrary, the emphasis is more on how an action contributes to the goal of life which is part of the natural order. Furthermore, in virtue ethics the virtues that contribute to someone's status as a full human being are varied, whereas in ETN the goal is much more unified. That is, in virtue ethics one can be virtuous in many ways; for example, one can be courageous, careful in spending money in certain situations while generous in others, and so on. Thus, there does not seem to be a single path toward achieving the highest goal as is the case in Buddhism or Stoicism. Achieving *eudaimonia* in virtue ethics is a matter of possessing certain character traits and acting on them, knowing when to utilize which trait or virtue in an appropriate way. On the contrary, in Buddhism, achieving *Nirvāṇa* is not merely a matter of possessing a number of virtues, but of there being one predominant virtue that is absolutely necessary for one to be able to attain the Goal, viz., knowledge of how things really are, or *prajñā*, which is usually translated as 'wisdom.' In Aristotle, the virtues are more or less equal to one another, but in Buddhism, as well as in Stoicism and Spinoza's ethics, wisdom (what Spinoza calls "the third (intuitive) kind of knowledge,"[5] which enables one to see how elements in Nature hang together in a unified way) is supremely necessary. Stoicism even goes so far as insisting that virtue is wisdom itself. This means that a person who possesses only the virtue of wisdom would eventually attain the status of the Stoic Sage, whereas one who possesses all other virtues but not wisdom alone would not be able to raise to that status. This does not mean, however, that in Buddhism, Stoicism and Spinoza's thinking, the other virtues are not important, but they are merely corollary to the development and perfection of wisdom. In this sense, wisdom is superior to ordinary kinds of knowledge such as perceptual knowledge, which is limited and "inadequate," in Spinoza's terms.[6] This point is also clear in Buddhism. Wisdom (*prajñā*) is crucially necessary

[5]Spinoza divides knowledge into three kinds. Roughly speaking, the first one consists of empirical knowledge, opinion and imagination. The second kind is rational knowledge, which is dependent on concepts and logic. The third kind, in Spinoza's words, "proceeds from an adequate idea of the absolute essence of certain attributes of God to the adequate knowledge of the essence of things" (Spinoza, *Ethics*, Book II, Proposition 40, Note 2). What Spinoza seems to have in mind here is that the third kind of knowledge results from a grasp of the essential relation that things have with God, or Nature. That is, the things' relation toward all other things, its place in the largest scheme of all things. This is the kind of knowledge that enables one to transcend all specificities and have an intuitive grasp of the whole.

[6]According to Spinoza, the only kind of knowledge that can be false is the first kind, which is empirical knowledge (Spinoza, *Ethics*, Book II, Proposition 41). This is because the second and third kind, being rational and intuitive respectively, consists of adequate ideas, which for Spinoza are ideas that, considered in themselves, have the qualities of a true idea. Perceptual knowledge always involves ideas that are inadequate because they are confused.

for the attainment of *Nirvāṇa*. In *The Path of Purification*, one of the most famous Buddhist texts dating from the fifth century CE, it is said that a monk who possesses virtue (*ṣila*), concentration (*samādhi*) and understanding or wisdom (*prajñā*) will be able to clear the path of purification and thereby reach the stage of *Nirvāṇa:*

> When a wise man, established well in virtue,
> develops consciousness and understanding,
> then as a bhikkhu [i.e. a monk] ardent and sagacious
> he succeeds in disentangling this tangle.[7]

Here, "disentangling the tangle" is a metaphor for clearing the way for the attainment of *Nirvāṇa*; this is accomplished through possessing virtue, concentration, which results in concentrated consciousness, and understanding, which is achieved through wisdom. Understanding here means that one has full knowledge of the way things are. For example, one knows that things are not permanent, but subject to decay and dissolution, and that everything is always interconnected. This understanding is thus comparable to the kind of rational understanding of nature in Stoic terms, and is also akin to Spinoza's notion of intuitive knowledge of God and God's Attributes.

Ethical teleological naturalism is also different from the standard modern ethical theories, particularly deontology and utilitarianism. The difference between ETN and Kantian ethics is clear. Kantian ethics does not emphasize metaphysics. The criterion of being good is based solely on the rational capability of an individual person—what one understands to be the maxim can be universalized as binding on every rational person. A distinctive feature of Kant's ethics is that it can function independently of metaphysics. No matter if there is a God or not, the maxim can always operate because it is not based on ontology, but on the rational thought and understanding of an individual human being. In fact, a human being is not even necessary for Kant's system to work. If we have a society of thinking, conscious and rational robots, then Kant's system would be at work because such robots also need to find a way to live together, which requires some forms of universalizable maxim. The difference between Kantian ethics and ETN is that the latter is metaphysical through and through. However, even in the society of robots, ETN could still function if the robots share the same goal, such as the ultimate goal of being born a robot, in the same way as Buddhists conceptualize the ultimate goal of being born a human being as eventually attaining *Nirvāṇa*. In this case, then, any action that facilitates such attainment would be considered in a positive light, and any action that does the opposite would be viewed in a negative light.

As for utilitarianism, at first it might seem that it is similar to ETN, but, in fact, the two views are radically different. The main difference between the two is that utilitarianism is a modern theory of ethics that dispenses with all the metaphysical underpinnings that give meaning to a theory that is directly based on ancient roots

[7] Buddhagosa, Bhadantacariya, *The Path of Purification (Visuddhimagga)*. 4th Ed. Transl. Bhikkhu Nanamoli. (Kandy, Sri Lanka: Buddhist Publication Society, 2010), p. 8. Available at http://www.accesstoinsight.org/lib/authors/nanamoli/PathofPurification2011.pdf

such as ETN. Thus, while both utilitarianism and ETN do emphasize the importance of consequences of one's action, the former does not specify that a good act must lead a person to attain one overriding goal of living: only following the own subjective preferences is enough. Utilitarianism emphasizes subjective preferences as a yardstick to measure how good or bad an action is. This does not mean that utilitarianism always endorses selfish acts—in fact any act that increases the overall amount of pleasure among the largest group of people does count as ethically positive. However, this means that it is ultimately the pleasure accruing to the "maximum number" that ultimately decides whether the act should be done or not. But, according to ETN, subjective preferences belonging to the maximum number are not a measure of their true interest. The main difference between the two here lies in what constitutes the true interest of a person. For Buddhism, it is a person's attainment of the own ultimate happiness, which is a result of a long process of training. But, for the utilitarian, one does not need any special training to attain one's own interest, since one's own subjective well-being is enough.

We have seen, then, that the ancient ethical system found in Buddhism in India and Stoicism in Europe does provide an alternative for the standard ethical theories in the West, namely deontological theory, utilitarianism and virtue ethics. This ancient system, however, is not free from internal debates either. In fact, this is to be expected from any healthy and complex ethical system. To illustrate this point and to show how it impacts normative argument development in Buddhist bioethics, in the following section, I will discuss the issue of euthanasia from Buddhist perspectives. Here 'euthanasia' is understood in a broad sense as including both the 'passive' and 'active' variety, where 'passive' means refraining from performing an action that uselessly prolongs the patient's life and 'active' actively doing something resulting in the patient's ending her life. Is euthanasia absolutely wrong in Buddhism? The answer is quite complex, as we shall see in the next section.

7.4 Different Views in Buddhist Ethics: How Should Euthanasia Be Understood?

Although Buddhist ethical theory is a kind of naturalism, it does not mean that there are necessarily no differences within the theory itself. Here, I analyze a debate between Rupert Gethin and Damien Keown, both of whom are well-respected scholars in Buddhist philosophy. Gethin is among those who regard euthanasia as categorically unethical from a Buddhist perspective, while Keown argues that Buddhism is more open and views euthanasia as more of a choice that people can make in order to relieve a loved one of unbearable pain and suffering. (The term 'suffering' here is used in its ordinary, non-technical sense, though there can be spiritual suffering too, which could be comparable to the Buddhist sense.)[8] Gethin's main argument is that the *Abhidhamma* specifically states that it is psychologically impossible for someone to kill with compassion. A common strand in Buddhist thought is that compassion is always required in any act that leads to positive results, as compassion means regard for the well-being of others, which always accompanies a reduction in selfishness and selfishness is a direct obstacle on the path toward Liberation. Thus, according to this view, killing always incurs negative results and is thus unwholesome and unethical. A consequence of this, according to Gethin's interpretation, is that compassion killing, i.e., the Buddhist version of mercy killing –killing or letting somebody die driven by compassion for that person–, is psychologically impossible. Because Gethin takes the *Abhidhamma* to be representative of Buddhism as a whole, Buddhism condemns euthanasia. According to Gethin, "the intention to kill is understood as exclusively unwholesome, and the possibility that it might ever be something wholesome prompted by thoughts of compassion is not countenanced."[9] The idea is that action always arises out of what is called a "motivational root." This is the underlying mental state that gives rise to an action, such as when one helps another person where the motivational root may be compassion toward that person, or when one hurts another person in which case the motivational root may be anger toward that particular person. In Gethin's interpretation of the *Abhidhamma*, killing another human being is always an act that harms that being. Hence, it is not possible to kill someone through compassion (meaning killing the person because of a regard for that person's own well-being) because any act that arises out of compassion must be one that relieves someone of her suffering. Since, for Gethin, killing cannot be a kind of action that relieves someone of her suffering because killing for him always involves violence and hatred, which is always inimical to compassion, killing cannot be done out of compassion. The outcome of this interpretation is clear: in Buddhism it is not possible to kill someone out of compassion, i.e., with the desire to relieve someone of their unbearable suffering. Therefore, euthanasia, which in Gethin's view includes forgoing treatment, is always immoral.

Against this, Damien Keown has argued that, while intentional killing is not allowed, it is ethically permissible to let die a patient who has no hope of getting better. Keown cites a passage in a commentary by Buddhagosa, one of the most respected commentators in Buddhism, where he says that when a patient has suffered for a long time with no prospect of getting better, it is legitimate for the nursing monks to deny food and medical care if it is certain that such help would merely

[8] See Rupert Gethin, "Can Killing a Living Being Ever Be an Act of Compassion? The analysis of the act of killing in the Abhidhamma and Pali Commentaries," *Journal of Buddhist Ethics* 11(2004), available at http://blogs.dickinson.edu/buddhistethics/2010/04/27/can-killing-a-living-being-ever-be-an-act-of-compassion-the-analysis-of-the-act-of-killing-in-the-abhidhamma-and-pali-commentaries/; Damien Keown, "On Compassionate Killing and the Abhidhamma's "Psychological Ethics," *Journal of Buddhist Ethics* 23(2016), available at http://blogs.dickinson.edu/buddhistethics/2016/02/29/is-compassionate-killing-psychologically-impossible/

[9] Rupert Gethin, "Can Killing a Living Being Ever Be an Act of Compassion?," p. 175.

prolong the suffering of the patient at the terminal stage.[10] This shows that, at least in the view of Buddhagosa, denial of food and medical care to a terminal patient may be allowed. Keown is careful in distinguishing between active euthanasia, where the intention is ending of the life of the patient, and this present care where it is pointless to continue medical treatment as death would merely be prolonged. The latter scenario would be passive euthanasia.[11]

Moreover, in a recent paper,[12] Keown takes aim at Gethin's view described above, arguing that the *Abhidhamma* has nothing against active or passive euthanasia per se, but the text merely argues that for an act to qualify as killing, it has to be rooted in negative motives such as hatred or delusion. Keown argues that killing need not be performed out of aversion in the mind. Thus, for Keown, it is possible to kill or end someone's life at the terminal stage out of compassion, whereas Gethin argues that this is never permissible in Buddhism.

All this shows that even among well-respected scholars of Buddhism, there are differences in how to interpret Buddhist texts as a way to find answers to pressing concerns of the modern world. It is not the aim of this present paper to resolve the debate between Keown and Gethin. Our concern is instead to reflect on the question how such differences could be understood and how this leads us to a theoretical understanding about the nature of normativity in Buddhist bioethics.

7.5 How to Deal with Differences in Buddhism?

As with other major intellectual and religious traditions, Buddhism contains a very large variety of diverse viewpoints, which engage in lively debates among one another. The debate between Keown and Gethin is a case in point that only illustrates how scholarly debates on the Buddhist view to ethical issues, which have been going on throughout the history of the religion, continue up to the present.[13] This makes us wonder whether within Buddhism itself there is a way to resolve doctrinal differences. To find the answer to that question, we have to go back to the early texts, which are assumed to be the closest we can get to the original words of the Buddha himself. Because – as we have seen – Buddhist ethics is an ethical teleological naturalism, the way to overcome doctrinal differences needs to lie within the purview of this theory. Roughly, this implies that a correct view within Buddhist philosophy must be one that is capable of leading believers to eventually achieve the ultimate Goal. A strand in Buddhist philosophy is that finding a practical way of

---

[10] Damien Keown, *Buddhist Ethics: A Very Short Introduction* (Oxford University Press, 2005), p. 112.

[11] Damien Keown, *Buddhist Ethics*, p. 112.

[12] Damien Keown, "On Compassionate Killing and the Abhidhamma's "Psychological Ethics."

[13] Even though Keown and Gethin are not practicing Buddhists, one can easily imagine that the debate between the two could readily occur among the practitioners themselves.

realizing the Goal is more important than knowing a truth, especially if that truth is not relevant to the realization of the Goal.

A well-known example is that of a disciple of the Buddha's who desired to know the answer to metaphysical questions such as whether the world is finite and infinite, and so on. The Buddha instead replied with the following parable. A man is being struck with an arrow. Here the pressing concern is to take the arrow out and treat the wound. But if the man who is struck keeps asking questions such as what kind of metal the arrowhead is made of, who shot it, from where, what kind of man shot him, from what caste, and so on, then it is likely that he will die before getting any answer.[14] The point of the story is that, in training oneself to attain Liberation, one should not pause and ask metaphysical questions such as whether the world is infinite or finite, or whether the Buddha will remain after death, or others of this kind. Asking all these will mire one in endless thoughts and arguments, which are not conducive to the Attainment of Liberation. Here, knowing all kinds of truth takes a back seat to performing all the practical tasks needed to attain Liberation. This does not imply that Buddhism is not interested in truth. Truth is supremely important especially when it does lead to the Goal. But, disputes as to whether a statement is true or not sometimes tends to lead attention away from the central concern, which is elimination of suffering. Thus, when there are differences in thoughts, the standard Buddhist response would be to see which one of the presented views does actually lead to the Goal. If one view does lead there while the other does not, then the first view should be preferred. But if the two views are equally good at attaining the Goal, then it is up to the individual to choose which one she should take.

So, with this in mind, let us return to the debate between Gethin and Keown. Which of these two views is more efficacious in bringing about the Goal? While searching an answer to this question, we have to bear in mind that, from the Buddhist perspective, the emphasis is on the Goal and not on the correctness of either of the two disputing views. From a philosophical perspective, this approach also makes sense, because in many philosophical disputes we simply do not know which view is the correct one. Buddhism is never fixated on a particular viewpoint. (We will expand on this important aspect of Buddhist metaethics in the next section.) According to Nāgārjuna, who lived in the second half of the second century and the first half of the third century CE and is considered one of the most influential Buddhist philosophers, the central aim of every Buddhist should even be to "relinquish all views."[15] A person who follows this recommendation is not attached to a particular view but instead focuses on how to arrive at Nirvāṇa as fast as possible. This implies that the content of a particular view is not as important as the practical

---

[14] The source of this story is the *Cula-Malunkyovada Sutta*. See "Cula-Malunkyovada Sutta: The Shorter Instructions to Malunkya," translated by Thanissaro Bhikkhu, available at http://www.accesstoinsight.org/tipitaka/mn/mn.063.than.html (retrieved October 8, 2017).

[15] Nāgārjuna, *The Fundamental Wisdom of the Middle Way*, Transl. Jay Garfield (Oxford University Press, 1995), Chapter XXVII, Verse 30.

goal of becoming released from all sufferings.[16] What is more important is whether or not the person holding on to a view has an attachment to it, because it is the attachment that leads a Buddhist away from the Goal. Now, Gethin's view is more fixed because he states that the *Abhidhamma* unequivocally says that killing cannot be done out of compassion. There seems to be no further room for maneuver in this interpretation. This shows that, in Gethin's argument, there is a degree of attachment to the defended view. Since such attachment is not conducive to attaining the Goal – as we have seen–, in comparison to Keown's argument, the spirit of Gethin's view is further from that of Buddhism, which is centered upon relinquishing any kind of attachment. In other words, the Buddhist position seems to favor a more sceptical stance—instead of being intent on finding the truth, one should be more open in admitting that one's own position could be wrong, becoming thereby disengaged from committing oneself to holding any particular viewpoint.

However, this conclusion should not be taken to imply that any content whatsoever is of equal value in bringing a Buddhist to *Nirvāṇa*, as long as there is no attachment to the content. The very first item of the Eightfold Path, that describes the Buddhist way leading up to *Nirvāṇa*, is *sammātiṭṭhi*, or Right View. A person needs to adhere to the right view in order to even get started on the Path. In order to understand this apparent paradox, we need to consider that the content of the Right View is always understood as the view that is effective in bringing about the Goal and that view will be, by definition, not extreme or fixed. In one of the Buddha's most important discourses on this topic, namely the *Kaccāna Gotta Sutta*, the Buddha explicitly says that the content of the Right View lies in the middle way between an assertion and its negation. When asked what constitutes the right view, the Buddha in this Sutta said that it consisted neither of affirming nor negating any categorical statements, but in maintaining that things are always dependent on one another: one thing follows from another. According to the Right View, something, a, can have a property, F, depending on certain factors, but in other situations it is possible that a does not have the property F because the factors may have changed. For example, a tomato is green in one condition, such as when it is not ripe, and red in another, such as when it is ripe.

Gethin's interpretation of the *Abhidhamma*, however, seems to presuppose that the statement of the *Abhidhamma* is fixed, namely that the motivational root of killing is always unwholesome (*akuṣala*). In comparison to that view, Keown's view is more flexible, and is thus more amenable to realizing the Goal than Gethin's. For Keown, under certain circumstances, such as when someone is suffering from a terminal stage of a disease, then those taking care of her can have a range of options at their disposal. They can decide to stop the ventilating machine, for example, an

[16] Scholars of Buddhism might object that Nāgārjuna belongs to the Mahayana tradition, which might not be representative of the whole of Buddhism. However, there is evidence that at least parts of Nāgārjuna's arguments do not deviate from the original teaching of the Buddha at all, and it has been argued that his work is an extended commentary on one of the Buddha's key teachings, the *Kaccana Gotta Sutta*. See Nāgārjuna, *The Philosophy of the Middle Way*, transl. David J. Kalupahana (Albany, NY: State University of New York Press, 1986).

option considered unacceptable by Gethin. This would be in line with Keown's argument that when the situation is really hopeless from a medical perspective, then an option would be that the caretakers stop prolonging life needlessly and withdraw life-support. It is important to note that, in this view, the caretakers do not need to do that. Withdrawing treatment is not mandatory. There is no ethical injunction in Buddhism that unambiguously instructs a person to do this or that in this type of situation. Buddhists have to rely more on their own individual judgment, which can vary according to situations. From this perspective, Buddhist's ethical theory is akin to virtue ethics in the West (though essentially it has more affinities with Stoicism, as we saw earlier). In virtue ethics, variability is due to an emphasis on character building. One who has practical wisdom knows that circumstances always vary, and one who has the practical wisdom knows how best to act accordingly. On the other hand, Buddhism, being more skeptical, starts from the acceptance that one should not become attached to any view and thus finds a way to accommodate varying circumstances in the most appropriate way. The important thing in Buddhist ethics is that no matter what decision is made, that decision has to be in line with the overall tenet of ethical teleological naturalism. That is, the action undertaken by the caregiver needs to ensure that the caregiver herself arrives one step closer to attainment of the Goal. This is what determines whether the action of the caregiver is good or bad. In more concrete terms, if it is an empirical fact that the action of the caregiver, such as turning off the ventilator, will result in the patient becoming released from the grip of unbearable pain and suffering where there is no hope to do so otherwise, then such action can qualify as having compassion as motivational root. This is so because compassion, or *karuṇā* in Buddhism, is the desire to relieve someone of her suffering. Thus, if the patient is suffering in such a way that the only way to relieve her suffering is to withhold the treatment, then out of compassion the caregiver should do it. Such an act need not be considered as 'killing' as defined in the *Abhidhamma*, whose usage by Gethin seems to presuppose that withdrawing life-support is always done out of unwholesome intentions or in a defiled state of mind (i.e., being beset by greed, anger or delusion).

It has to be noted that the foregoing is only one way of looking at how to resolve differences between Buddhist approaches. Another way, which is much simpler, starts from the observation that Gethin and Keown refer to different passages in the Buddhist scriptures to support their claims. A large and enduring tradition such as Buddhism certainly contains within itself views that are divergent from one another. This comes as no surprise given the long history of Buddhism and the sheer number of scholars and monks who have worked on the Scripture for millennia. Even the *Abhidhamma*, though recognized as one of the Three Baskets commonly thought to represent the original teaching of the Buddha, clearly contains later additions, such as passages detailing how the Buddha's view has been distorted in various ways after his death. Obviously, these cannot be the Buddha's own original words.[17]

[17] See, for example, Erich Frauwallner, *Studies in Abhidharma Literature and the Origins of Buddhist Philosophical Systems* (Albany, NY: State University of New York Press, 1995).

Keown largely relies on the commentary by Buddhagosa, a scholar living centuries after the Buddha himself. Thus, we could look at the euthanasia debate as a conflict among the Buddhist texts themselves. It is, then, the task of the interpreter to reconcile these differences. It is apparent that there are different authors, one who composed the *Abhidhamma* text cited by Gethin, and Buddhagosa himself. These authors are credited as being highly authoritative within Buddhism, but when their authorities collide, later scholars can provide their own interpretation trying to reconcile the two. One approach would be to look at different contexts in which the *Abhidhamma* and the commentary by Buddhagosa were written. For example, the scholar could look at the *Abhidhamma* text and argue that the idea of 'compassion killing' or euthanasia could not have arisen in the mind of the author of the relevant *Abhidhamma* text because in the latter's context mercy killing was not a predominant issue. What frequently did happen was manslaughter and murder. Therefore, when Buddhagosa analyzed the motivational roots that gave rise to killing, he primarily thought of cases that are the consequence of defiled states of mind, namely greed, anger and delusion. The use of ventilators and other sophisticated technologies to keep patients alive were not available in the Buddha's time, or in the time of the *Abhidhamma*'s author, so neither could have imagined that there could be situations where letting death come naturally would be a serious ethical issue.[18]

7.6 Buddhist Pragmatism

The idea that following norms leads to the desired Goal shows that Buddhist metaethics is pragmatic in nature. Here, pragmatism in following norms just means that a person follows a norm because it leads to what he or she wants. In an earlier paper,[19] I argued that when people have to choose a universal norm from among a set of competing norms proposed by many cultures, they should choose the norm that serves the interests of everyone. Yet, that is only possible if the interests of everyone converge.[20] Thus, if a set of norms has become a more or less global standard, then a culture that already has a different set needs to adjust its own norms in order to be able to interact with the other cultures. In the contemporary world, the prevailing norms contain values such as respect for human rights and individual

[18] For example, my colleague Somparn Promta is well known for his interpretation of Buddhist ethics as one that admits of necessary evils, i.e., wrong action that is necessary for the well-functioning of society (Somparn Promta, *Buddhism and Ethical Problems* (Bangkok: Putthachart Press, 1992 [in Thai]). In the section on euthanasia, he also argues that euthanasia is permissible if it is necessary for society and for relieving the pain of the patient, even though according to the strict rule of Buddhism it is wrong.

[19] Soraj Hongladarom, "Intercultural Information Ethics: A Pragmatic Consideration," in Matthew Kelly and Jared Bielby, eds., *Information Cultures in the Digital Age* (Springer, 2016), pp. 191–206.

[20] This view is a form of consequentialism but in that paper I do not claim that the view presented there is Buddhist, because Buddhist ethics is based on a metaphysics, as we have seen.

preferences. Thus, cultures that hold onto norms contradicting the global standard, will need to adjust their own norms in order to join in the international community. The goal in this case is to be able to join the international community, and the means, of course, is adoption of the global norms.[21] This does not mean that the global norms are unnegotiable. There must be a constant effort involving all stakeholders to negotiate and adjust the global norms. In my earlier paper, I emphasize that it is not useful for people belonging to a particular culture to adamantly insist on their own particular viewpoint which they may believe to be right. Working with others in dialogue presupposes that one is willing to question and even give up the own established convictions along the way. So, instead of focusing on which one from among conflicting norms is the true one, the parties should seek to find which values from all these norms enable the parties themselves to realize their goals most effectively, in this way emphasizing the goals over the norms.[22]

I did not reveal in that paper that the inspiration of the ideas presented there was entirely Buddhist. I call this view Buddhist pragmatism. Here, pragmatism dovetails with ethical teleological naturalism, in which values are natural properties. Achieving those values is more a matter of realizing those natural properties empirically and, when there are competing sets of values, seeing which sets achieve the properties more effectively, which is the hallmark of pragmatism.[23] As we have seen above, in Buddhist pragmatism, truth takes a back seat to efficacy. By adopting this approach, Buddhism finds itself in agreement with the ancient Greek skeptic Pyrrho (c. 360-c. 270 BCE), who argued that people should not adhere to any particular view, because that would imply attachment. In Pyrrho's view, such attachment should be avoided as it leads to the condition of being easily perturbed, which, for Pyrrho, is the opposite of the desired condition of being calm and free from passions. That condition is called *ataraxia.* There is a recent work claiming that Pyrrho received this view from Buddhist monks in India when he accompanied the army of Alexander the Great there.[24] Whether or not this is the case, Pyrrho's view seems close to that of Buddhism. For Pyrrho, the goal is to achieve ataraxia through avoidance of judgment on philosophical matters. For the Buddha, the goal is to achieve *Nirvāṇa* through following the "middle way," which urges Buddhists not to judge whether things either are or are not. The upshot of both the Buddhist view as well as that of Pyrrho is that the truth of a claim's content is not the most important matter. It is more important to withdraw from judging whether a claim is either true or false and instead focus on the positive effect on one's own mind of such withdrawal.

[21] Soraj Hongladarom, "Intercultural Information Ethics: A Pragmatic Consideration."

[22] Soraj Hongladarom, "Intercultural Information Ethics: A Pragmatic Consideration," pp. 196–197.

[23] In this way values are both natural properties and ways to achieve further properties. That is, values, as natural properties themselves, in this context, act as a means by which further natural properties can be achieved. Being diligent is a property which is valued positively; it usually brings about being successful, which is also valued positively.

[24] Christopher Beckwith, *The Greek Buddha: Pyrrho's Encounter with Early Buddhism in Central Asia* (Princeton, NJ: Princeton University Press, 2017.

What makes a claim ultimately valuable depends on whether one completely refrains from judging whether the claim is right or wrong, which is necessary for destroying attachment to it. And, it is such destruction of attachment that is a key factor in attaining *Nirvāṇa* in Buddhism or *ataraxia* in Pyrrho's philosophy.

If we carry this Buddhist philosophical tenet over to the problem of finding a universal set of norms, we have to conclude that we cannot determine whether a norm is universal or not by examining its content. In fact, no norm can be universal according to Buddhism if by being universal we mean that the norm is applicable at all times and all places. A norm can be universal only when the issues or contexts to which the norm is applied are everywhere and always exactly the same, which is impossible. However, the fact that there is no universal norm does not mean that there is no way of ranking norms as more or less valuable or desirable. This can be straightforwardly done by assessing how much or how little a norm contributes to the Goal.

When we, now, apply this Buddhist pragmatism to the issue of euthanasia, we have to assess to what extent the proposed norms would lead to cessation of suffering (*dukkha*). In the Buddha's times, euthanasia was not a major issue because there was no technology to prolong life indefinitely. Given the fact that we do have such technology now, the task for Buddhists in today's world is finding out how to interpret the notion of 'suffering' so that it can be used to assess whether or not terminally ill and dying patients do suffer in the Buddhist sense. To do that, it will be necessary to analyze the notion of meaning in life. From that perspective, we can wonder what the difference is between someone who has already died and someone who is being artificially propped up by a machine, has lost consciousness, and may be enduring pain. If the difference is only that the latter is still breathing (albeit through the machine) and the first is not, then it seems that withdrawing life support systems would be preferable because it leads to at least elimination of the pain that the patient is suffering from. This conclusion perfectly illustrates Buddhist pragmatism. It does not condemn or endorse euthanasia per se. In other words, for Buddhism euthanasia of any kind is neither good nor bad in itself. What makes it either good or bad is the intention behind the action. If there can be a compassionate act of letting someone die who is suffering from a terminal illness and is in unbearable pain, then such an action may not be wrong from the perspective of Buddhist thought. If Buddhists do not want their religion to be a frozen system inherited from the past, they have to find ways to make the teaching relevant by figuring out how to respond to cases involving new technologies that masters in the past had no means of even imagining. They need to do this without destroying the spirit of Buddhism itself.

However, the application of Buddhist pragmatism to the case of euthanasia does not end there. An objection to the above reasoning could be that the cessation of suffering should not be assessed on the basis of suffering of other people, i.e. the patients, but the suffering of the healthcare provider administering euthanasia. In addition, a key factor in determining whether an action is good or bad is the intention (*cetanā*) of the act. If the intention or motivation is good (*kuṣala*) then the action is good, and if not then it is bad. This basic understanding of the Buddhist teaching is reflected in Gethin's interpretation of the *Abhidhamma,* as we have seen.

The desire to relieve others of suffering has to be rooted in a good intention—i.e. the intention to see that others do not suffer. In order to be able to do this, Buddhists have to cease looking after their own self-interest and put others above themselves. This is the core of the Buddhist compassion: the desire to relieve others of suffering as well as actual practice to realize that desire. In desiring that others cease to suffer, a person is motivated by altruism and compassion. From that perspective, desiring that others cease to suffer is in line with the own self-interest because altruism and compassion are essential values towards the Goal. So, ironically, in caring for others, people do indeed take care of themselves. On the other hand, cherishing the own ego is always inimical to realizing the Goal. Therefore, a practice that seeks to eliminate attachment to the ego is always good as it contributes to the Goal. In caring for others, then, a Buddhist has an opportunity to practice compassion, which leads to the betterment of oneself in the end.

In the context of the use of modern medical technologies where the patient is in irreversible coma and the use of life-sustaining equipment merely prolongs suffering and pain, it is of utmost importance to determine whether or not in undertaking euthanasia the desire is to end suffering. Using that criterion, it becomes possible to argue that undertaking euthanasia in this context is not the same as killing as mentioned in the *Abhidhamma*. The author of the *Abhidhamma* may only have thought of killing in the usual context of ending a person's life where that life is not devoid of meaning, as may be the case when a patient is tied up to a ventilator or other sophisticated life prolonging devices. However, such an argument is only possible when Buddhists look at their religion in a broader context and try to adapt traditional teaching to changing times and without rigidly adhering to what is literally said in the text.

7.7 Conclusion

I have argued that Buddhist metaethics is a kind of ethical teleological naturalism: a view that positive ethical values are those that are efficacious in bringing about the desired goal, which in Buddhism invariably is *Nirvāṇa*. Moreover, Buddhist metaethics is also a kind of pragmatism when it comes to the question of choosing from among competing sets of values. Those values that are more effective in bringing about the Goal are to be preferred. The reason for the preference towards the Goal is not that certain sets accords with human rationality as in the modern theory such as Kant's, nor that those sets bring about pure subjective preferences as in utilitarianism. I have argued that Buddhism rejects these because they are not based on a metaphysical conception of human nature and suffering (*dukkha*). Buddhist metaethics does find itself allied with ancient Western philosophies such as Stoicism and Pyrrhonic skepticism, as well as the more neglected 'modern' ethical theory of Spinoza. It, also, has some affinities with Aristotelian virtue ethics in aiming to achieve the Good Life, which can be conceived in terms of either *Nirvāṇa* or *eudaimonia*, but it differs from Aristotle's thinking in that the Buddhist metaethics does not see that the virtues are the first key factors in achieving the Goal. In other words, the only ultimately relevant virtue in Buddhism is wisdom, and here is where Buddhist ethics finds itself more aligned with Stoicism, which identifies virtue with knowledge or wisdom.

Buddhist metaethics is relevant to modern bioethical debates in that it provides a theory of normativity that allows for a metaphysical or natural conception of the human body and mind and dovetails ethical values and judgments into the human nature. I also discussed a recent debate by two Buddhist scholars, Rupert Gethin and Damien Keown, on the subject of the ethical value of euthanasia, and found that Keown's more flexible position is more in line with the spirit of Buddhism.

In conclusion, then, in the current atmosphere where bioethical research is still largely dominated by the modern theories of normativity derived from Kantianism or utilitarianism, Buddhism offers an interesting alternative, one which brings us back to the realization that metaphysics (broadly construed as a description of human nature) is still relevant to ethics, which implies that empirical method is still relevant to ethical deliberation. That is, what one really knows about the nature of the world is deeply relevant to what one conceives to be either good or bad, right or wrong. Buddhist theory is not even one that is foreign to the mindset of an average Westerner, because it has many affinities with ancient theories in the West itself, namely Stoicism and Pyrrhonic skepticism, as well as a modern philosophy, such as Spinoza's. It may be time now to return to these ancient theories to find new sources of insight.[25]

Suggested Readings

Buddhist Ethics in General

Cea, A.V.D. 2004. The Criteria of Goodness in the Pāli Nikāyas and the Nature of Buddhist Ethics. *Journal of Buddhist Ethics* 11. Available at http://blogs.dickinson.edu/buddhistethics/2010/04/27/the-criteria-of-goodness-in-the-pali-nikayas-and-the-nature-of-buddhist-ethics/

Hallisey, C. 1996. Ethical Particularism in Theravada Buddhism. *Journal of Buddhist Ethics* 3: 32–43. Available at http://blogs.dickinson.edu/buddhistethics/files/2010/04/hallisey.pdf

Harvey, Peter. 2000. *An Introduction to Buddhist Ethics: Foundations, Values and Issues*. Cambridge: Cambridge University Press.

Keown, Damien. 1992. *The Nature of Buddhist Ethics*. New York: Palgrave Macmillan.

[25] I would like to thank Henk ten Have for inviting me to the very fruitful conference on Normativity in Bioethics in Leuven in May 2016. My thanks also go to Joris Gielen for his superb organization. Research for this paper is partially supported by a grant from Chulalongkorn University, under the Ratchadapisek Fund for Chulalongkorn Research Fellows, 2015–2018.

Buddhist Bioethics

Fujii, Masao. 1991. Buddhism and Bioethics. In *Bioethics Yearbook*, 61–68. Netherlands: Springer.
Gethin, Rupert. 2004. Can Killing a Living Being Ever Be an Act of Compassion? The Analysis of the Act of Killing in the *Abhidhamma* and *Pali* Commentaries. *Journal of Buddhist Ethics* 11. Available at http://blogs.dickinson.edu/buddhistethics/2010/04/27/can-killing-a-living-being-ever-be-an-act-of-compassion-the-analysis-of-the-act-of-killing-in-the-abhidhamma-and-pali-commentaries/
Hongladarom, Soraj. 2009. Privacy, the Individual and Genetic Information: A Buddhist Perspective. *Bioethics* 23 (7): 403–412.
Hughes, James. 2006. Buddhist Bioethics. In *Principles of Health Care Ethics,* ed. Richard E. Ashcroft, Angus Dawson, Heather Draper, and John R. McMillan, 127–133. Chichester: Wiley.
Keown, Damien. 1995. *Buddhism & Bioethics*. New York: St. Martin's Press.
Keown, Damien. 2005. End of life: The Buddhist View. *The Lancet* 366 (9489): 952.
Keown, Damien. 2016. On Compassionate Killing and the Abhidhamma's "Psychological Ethics". *Journal of Buddhist Ethics* 23. Available at http://blogs.dickinson.edu/buddhistethics/2016/02/29/is-compassionate-killing-psychologically-impossible/
Schlieter, Jens. 2014a. Some Aspects of the Buddhist Assessment of Human Cloning. In *Human Dignity and Human Cloning*, ed. S. Vöneky and R. Wolfrum. Dordrecht: Springer.
Schlieter, Jens. 2014b. Endure, Adapt, or Overcome? The Concept of Suffering in Buddhist Bioethics. In *Suffering and Bioethics*, ed. Ronald M. Green and Nathan J. Palpant, 309–336. New York: Oxford University Press.

Part II
Normativity at Crossroads

Chapter 8
Introduction to Part II: Normativity at Crossroads

Joris Gielen

Keywords Normativity · Bioethics · Empirical research · Secularism · Pluralism · Narrative · Ethics consultant

Even though religious bioethicists have tried to find ways to develop normative bioethical arguments amidst the diversity of their respective religious traditions, the various chapters of the first part of the book may still leave the impression that a study of religious bioethics actually leads to the conclusion that within a global context bioethical arguments cannot be normative for everyone, but only for those who adhere to particular traditions. In this part, we assess the validity of that statement and explore ways to talk about normativity in a global context and implement normative discourse in practice.

The analysis of normativity in bioethical reasoning in religious traditions that was presented in the first part can, in a way, be seen as empirical data. The authors of these chapters have described various religious approaches to normative bioethics. Do empirical observations, then, show us that normative bioethics is not possible because the various approaches to normative bioethics are so radically different? In Chap. 9, Rouven Porz and Pascal Borry argue the opposite. They are of the opinion that empirical research can effectively contribute to normative bioethics. They present a critical applied ethics model in which empirical data and normative theories interact.

Of course, the model that Porz and Borry have developed is particularly aimed at sociological and anthropological research that studies contexts and topics that are related to bioethics, such as attitudes of healthcare professionals to controversial treatments or the experiences of people who are undergoing medical treatment. As Porz and Borry show, such data can contribute to the development of normative arguments, but are not normative on their own. The fact that such data, which may challenge normative bioethical arguments can, nevertheless, be integrated in such

J. Gielen (✉)
Center for Healthcare Ethics, Duquesne University, Pittsburgh, PA, USA
e-mail: gielenj@duq.edu

J. Gielen (ed.), *Dealing with Bioethical Issues in a Globalized World*, Advancing Global Bioethics 14, https://doi.org/10.1007/978-3-030-30432-4_8

arguments, provides interesting perspectives for the sustainability of normative bioethics in a globalized world. In global bioethics, we see a similar dialectical process at work. In Chap. 10, Henk ten Have develops the idea that global bioethics transcends the individual perspective and has a social-ethical nature. It, therefore, needs some kind of global framework. In global bioethics, there is a framework of universal principles. These principles can be upheld – ten Have argues – by seeing normativity as dialectics. Global bioethics assumes that there are principles on which there can be broad agreement among people belonging to different cultures and traditions. These principles, however, need to be contextualized due to the particularity of each health-setting. At the same time, these local applications inform the interpretation and formulation of the principles.

A similar model may also be helpful to practically deal with bioethical conflict in our contemporary societies that are simultaneously secular and pluralistic. The argument presented in Chap. 11 is that, in such a context, moral relativism may look appealing, but it is both untenable and unpractical. The theory of common morality shows a theoretical and practical way out of this problem. The common morality provides the basis for dialogue and understanding when ethical conflict arises in healthcare, while diversity and compromise are possible at the level of particular moralities.

However, such dialogue is not a given in healthcare. It requires healthcare providers to pay attention to the patients' culturally and socially embedded narratives. In Chap. 12, Fahmida Hossain argues that healthcare organizations have a duty to orient themselves towards these narratives if they want to overcome bias and inequality. Fahmida Hossain calls that guiding focus on the patients' narratives "narrative authority". Narrative authority makes healthcare organizations culturally competent so that they can thrive and resolve ethical conflict in this era of globalization.

The resolution of ethical conflict will often, also, require the involvement of ethics professionals: ethics consultants. However, we can wonder why patients, their relatives, and clinicians would accept the involvement of ethics consultants. Considering the diversity of contemporary society, they may assume that ethics consultants may be biased towards ethical positions that differ from those aligning with their own worldview or religion. For these consultants to function effectively within the current healthcare context and facilitate ethics, they need some kind of authority. In Chap. 13, Kenneth Kipnis argues that ethics consultants have ethics expertise that enables them to provide guidance in complex ethical cases in healthcare. Within the context of the argument that is developed in this book, it can be argued that this particular ethics expertise includes attentiveness to narrative authority. When the consultant's ethics expertise integrates narrative authority she will be able to facilitate ethics in healthcare in a globalized world.

Chapter 9
Combining Empirical Data and Normativity: Possible or Not?

Rouven Porz and Pascal Borry

Abstract Researchers in the field of bioethics have increasingly been confronted with empirical research on bioethical questions. This development has raised the question as to the extent to which empirical contributions both relate to normative approaches in bioethics and aspire to achieve normative statements about the topics under study. Therefore, in this article we aim to reflect on the contribution of empirical research to normative ethical reflection and decision-making. Firstly, we describe narrative. Narratives allow us to reflect on moral dilemmas in ways that are different to what a theoretical approach allows. Secondly, we describe the different contributions that empirical research could have for ethical reflection and decision-making, as is described in the literature. Thirdly, we describe the critical applied ethics model, which is a theoretical approach for viewing the interaction between empirical data and normative theories as a two-way relationship.

Keywords Bioethics · Empirical ethics · Methodology · Normativity

9.1 Introduction

Researchers in the field of bioethics have increasingly been confronted with empirical research on bioethical questions (Borry et al. 2005; Sugarman 2004). The actual increase of empirical contributions in the field of bioethics has been intensively commentated and has been labelled (with sometimes slightly different interpretations) as "empirical research in bioethics" (Sugarman 2004), "empirical-ethical

R. Porz
Medical Ethics, Medical Directorate, Inselspital, Bern University Hospital, University of Bern, Bern, Switzerland
e-mail: Rouven.Porz@insel.ch

P. Borry (✉)
Centre for Biomedical Ethics and Law, Department of Public Health and Primary Care, KU Leuven, Leuven, Belgium
e-mail: Pascal.Borry@kuleuven.be

J. Gielen (ed.), *Dealing with Bioethical Issues in a Globalized World*, Advancing Global Bioethics 14, https://doi.org/10.1007/978-3-030-30432-4_9

research" (van der Scheer et al. 2004), "social sciences perspectives on bioethics" (Weisz 1990), "transdisciplinary research in bioethics"; "sociology of bioethics"; "evidence-based ethics"; "cross-cultural bioethics" and "ethics-related empirical research" (Miller 2002). This demonstrates that a broad perspective is important in this area. As mentioned by Kon, much empirical work in the field of bioethics

> has been conducted by individuals who do not consider themselves bioethicists, who do not consider their research to be research in the field of bioethics, and who publish their work in journals that are not generally considered bioethics journals. This finding points to the broad nature of research that may influence bioethics and potentially may lead to changes in our ethical norms. Further, many who are considered leaders in bioethics do not consider themselves to be bioethicists per se, but rather clinicians or researchers whose work is related to bioethics. (2009)

Therefore, when we refer to empirical research in bioethics we mean

> the systematic investigation of a particular problem in the field of bioethics by using a research methodology (such as a qualitative or a quantitative approach) that has methodological roots in the social sciences with a view to generating, analysing and interpreting reliable and valid information or data about the total experience of all members of the society or the group under study. (Leget and Borry 2010)

This development has raised the question as to the extent to which empirical contributions both relate to normative approaches in bioethics and aspire to achieve normative statements about the topics under study. Fundamentally, the question arises as to the extent descriptions of what *is* the case help in defining what *should* be the case. This distinction goes back to the divide between normative and non-normative approaches. The goal of normative ethics is to set out in a rational, coherent, anti-dogmatic and argumentative way what people should do (van Willigenburg et al. 1993). In contrast, descriptive ethics has been called the factual investigation of behaviour, norms, rules attitudes and views occurring or existing in society. In focusing on the strict distinction between *is* and *ought*, empirical research has often been considered to provide "just the facts" (Nelson 2000), while normative ethics focuses on the understanding of value issues and the development of arguments in order to advance appropriate answers and positions.

Therefore, in this article we aim to reflect on the contribution of empirical research to normative ethical reflection and decision-making. Firstly, we will start this contribution with a narrative. Narratives allow us to reflect on moral dilemmas in ways that are different to what a theoretical approach allows. They help us to explore moral issues and engage in reflection (Nelson 2014). Secondly, we will describe the different contributions that empirical research could have for ethical reflection and decision-making, as is described in the literature. Thirdly, we will describe the critical applied ethics model, which is a theoretical approach for viewing the interaction between empirical data and normative theories as a two-way relationship.

9.2 Analogy – Snorkelling in the Rhine

We will start the reflection on the contribution of empirical research to bioethics with a narrative by a co-author of this contribution (Porz 2008).

> In Basel, Switzerland, it is a recognised and commonly practiced sport, to go swimming in the Rhine. For newcomers this can be a somewhat disconcerting spectacle, as several hundred people let themselves be carried away by the current with jolly facial expressions. I had never seen something like this before. All age-groups are represented in the water, except for very young children. I am not speaking of 'going swimming' here, but of 'being swept along the river'. The city Basel has even marked special areas in the river, in which 'the sweeping' is permitted. This is an important note, seeing as swimming in the Rhine is virtually being promoted by the city. As a newcomer I placed myself right at the water's edge, thinking: 'No, you should not do this, it is too dangerous – and the Rhine is too dirty'. This would have been a normative statement with the justification of a bodily danger – and a reference to my ecological attitude. Danger is found not only in the current, but also from the freighters and tourist boats the Rhine is frequented by. The swimmers find themselves on eye level with the passing ships. I did not express my indignation, but spent my first summer in Basel watching the swimmers from the esplanade of the Rhine. Through this participatory observation, I realised two things: nobody swims alone, and everyone carries a waterproof bag with them. This bag hangs on the back of each swimmer and does not only include dry clothes and towels for when they arrive downstream and want to exit the water, but they can also hold onto their air-filled bag when threatened with drowning. A first reassurance for me – an observation, just as if I were an anthropologist.
>
> In my second year in Basel, I crossed the Rhine several times in a small ferry and could therefore look at the swimmers from this vantage point. Now I was on eye level, I was closer to them – but I still wasn't in the water. I was still tempted to say: you should not do this, it is too dangerous (my normative baseline assumption!). I considered getting into the Rhine with diving equipment. With a wetsuit and respirator I could have watched the swimmers from below. Could I then still permit myself to make a statement about their 'dangerous' and, for me, elementarily 'incomprehensible' hobby? There was only one way to develop a justifiable opinion: I had to enter the river. I had to have the experience myself, by virtually plunging into my own empiricism.
>
> Thereto, I decided at a minimum to put on my snorkelling goggles. This way I had the option of seeing above, and also below, the water level (through which I could, for instance, see hidden obstacles like big stones. Please note: I was predominantly scared of bumping into something that I could not see at first sight).
>
> Now: the experience of swimming in the Rhine was fantastic and barely describable in words. One finds oneself in the current with the whole body, which encompasses one. One gets pulled along gently and one can get back to the shore at any time. I could even get into conversations with the other swimmers. Most of them had practiced swimming since their youth, had developed specific rituals and evidently enjoyed talking about it. One cannot deny that I remained an outsider. I was still wearing my snorkelling goggles, looked below the water, head up, head down, afraid of big stones down there, then once turned around to the tourists who, just like me in the year before, strolled around the esplanade in astonishment. And then it happened: "peeehnnng" with great momentum, my head bumped into a Rhinebuoy. It was clearly visible in the water, I just did not pay attention. Very dazed, I held onto my waterproof bag, which I had luckily brought on the basis of my participatory observation, and dragged my exhausted self to the shore. I had never anticipated that one could bump his head against a, from 1000 meters visible, fluorescent buoy. The thought would never have crossed my mind and it did hurt painfully.
>
> A few weeks later, friends from Germany visited me. We intended to go swimming in the Rhine together but it took an immense amount of persuasion to encourage my friends

into the river. Already in swim shorts, with snorkel and snorkelling goggles, I told them: "you can swim here, it is not as dangerous as it looks – but you have watch out for the buoys". Without my previous personal involvement, I would not have been able to form this normative judgment. Of course, I will never be a true local, I will remain the outsider with snorkelling goggles, but I was as close as I could possibly get.

9.3 Models for Relating Empirical Research to Normative Ethics

This story can easily be viewed as a reflection on the value of empirical research in bioethics. The snorkeler is the enquiring researcher and the swimmers in the Rhine are the societal group to be investigated. Some societal developments are – just like swimming in the Rhine – so novel to some observers, that one is easily tempted to shout: "Stop it – it is too dangerous". But what is this judgment based on? With empirical research one can leave the heights of the Rhine esplanade, without taking the detour with the ferry, by departing from the depths of philosophical reflection. One goes to the surface of the water, to the point where the development actually happens. This does not mean that one cannot take a glimpse through ethical snorkelling goggles at moral depths, yet one attempts to balance oneself, head up, head down. Simultaneously one tries to comprehend why people act the way they do in their daily (or professional) lives. One has the position that the people to be studied are specialists for their own lives and discovers values and facts that one had never thought of before (the buoy!). One notices what people understand as values and what they think of as facts. Naturally, honesty is still required as is the ability to reflect on the situation so as to compare the discovered 'data' with ones' own basic normative assumptions, or to place it in a relationship with ethical theories. The lived experiences alone cannot dictate what is morally right, but at least a dilemma can be discussed from the perspective of a lived experience. Those who actually experience the water might find it easier to align their moral conclusions in relation to the real world of the water.

Analogies are of course never precise or complete, and they can easily be criticized. Nevertheless, they allow us to reflect on the relationship between approaches that try to explain and understand reality, and other approaches that normatively judge reality. The way that the experience described above could influence normative views will of course depend on the inherent position taken on the normative value of this experience. Molewijk et al. described, in an important contribution, five possible models by which empirical research can be related to normative ethics: prescriptive applied ethics, theorist ethics, critical applied ethics, particularist ethics and integrated empirical ethics (2004).

In the first approach (prescriptive applied ethics), "empirical data will never cause substantial adjustments of a moral theory". Empirical data play an instrumental role, i.e. in understanding social practices or the practical consequences of certain behaviour. Moral theories would be considered the "final arbiter." As expressed by Molewijk et al. "In this approach, a moral theory does not lose its legitimacy when it does not concur with the morality of a certain social practice" (2004). In the second approach (theorist ethics), empirical data receives a slightly more important role. Empirical data is considered to be able to refine a moral position or a moral theory. Although values, norms, and moral theories remain the final arbiter, adjustments of moral theories or normative positions are possible. This approach is "cognizant of empirical results" (2004).

The third position described by Molewijk and his colleagues is *critical applied ethics*. This position allows for a two-way relationship between empirical data and normative theories: a social practice can and should be judged by both the gathering of empirical data and by normative ethics. Both methods of looking at a social practice are open to readjustment and refinement and each can perform this role for the other. In previous contributions (Leget and Borry 2010; Leget et al. 2009), we described this position as a five-stage model, in which empirical and normative approaches are distinguished as two independent entities that interact in a critical way. We will describe this five-stage model in the last section of this article.

The fourth approach (particularism) sees the morality of a social practice as "the one and only moral arbiter" (2004). As a consequence, "there is no interaction between empirical data and moral theory" and there is no wish to "connect the morality of a social practice with a moral theory" (2004). This approach gives so much weight to *social context* that *moral theory* no longer performs a critical role. General acceptance of a practice by a community suffices as ethical justification of that practice.

The fifth position described by Molewijk and his colleagues is their preferred position. Labelled *integrated empirical ethics* (IEE), it is an intensified version of *critical applied ethics,* an ongoing dialogue between descriptive social scientists and ethicists in which the distinction between fact and value eventually disappears. As Molewijk et al. rightly point out, facts always contain and generate values, and values are based on facts (2004). In their view, this sociological truth demands a theoretical hybridization of the normative and empirical disciplines, in which the cores of the two disciplines are combined to form a new field of research (2004).

9.4 Identifying the Contribution of Empirical Research to Normative Decision-Making, and Vice-Versa

The third position in the contribution of Molewijk et al. has the advantage of continuing to distinguish the normative and the empirical as two independent entities that are intrinsically linked to each other. In a previous contribution (Borry et al. 2004) we described a few characteristics of ethical decision-making. Firstly, "an ethical problem expects to be answered in terms of good, evil, right or wrong" (2004). You expect a normative answer to a particular problem. Secondly, "ethical decision-making is not a matter of absolutely right or absolutely wrong, but a careful weighing of relevant norms and values. Moral truth remains the goal, but is unattainable as such" (2004). Thirdly, ethical decision-making cannot be merely a subjective matter, but can only be regarded as acceptable if a decision is supported by strong arguments. Fourthly, ethical decision-making has to be consensus-driven and make efforts to integrate the views of various stakeholders. In the process of ethical clarification and decision-making, various steps can be identified. We described earlier in much more detail a five-step approach (Leget and Borry 2010; Leget et al. 2009) which includes: (Sect. 9.4.1) the determination of the problem, (Sect. 9.4.2) the description of the problem, (Sect. 9.4.3) the study of effects and alternatives, (Sect. 9.4.4) normative weighing and (Sect. 9.4.5) the evaluation of the effects of a decision. At every step, empirical and normative approaches can mutually assist each other.

9.4.1 *The Determination of the Problem*

As expressed by De Vries and Subedi, "morality is socially situated. The decision to become 'ethical' is not made in a vacuum: Ideas about right and wrong, proper and improper, are shaped by social context" (1998). Framing a problem as an ethical problem can be done through either an empirical or normative approach. Empirical research can help identify ethical problems (Braddock 3rd, 1994; Brody 1993) and it can lead to increased attention for subjects deserving of ethical reflection, or issues that only receive minimal ethical consideration (Fulford 1994). In contrast, normative perspectives can critically challenge existing norms or bring sensitivity to issues that are currently not perceived as being ethical.

9.4.2 *Describing the Problem*

As Sulmasy and Sugarman have noted, a number of empirical research tasks undertaken in bioethics are "purely descriptive studies of what human beings believe about morality, how their beliefs change with time, and how they behave in situations of moral concern" (2001). Although the techniques used are descriptive, and the conclusions do not have immediate normative implications, these studies are "interesting precisely because they illuminate human responses to normative questions" (2001). They might introduce "facts not already being considered in reaching normative conclusions, thereby better informing this work" (2001). Normativity is socially situated: empirical studies can therefore help situate normative concepts and ethical positions in their broad social context (Ackerman 1980; Fox and de Vries 1998).

9.4.3 *Effects and Alternatives*

Normative decision-making might create changes in reality. It would be inappropriate not to take consequences into account when making an ethical decision. As Brody pointed out:

> even if one is not a consequentialist, one can at least agree that consequences are morally relevant and are part of the basis for evaluating policies. In this way, then, discoveries about what *is* the case are relevant for deciding what *ought* to be the case. (1993)

Empirical research can help clarify the consequences of a decision or policy. It might help elucidate so-called slippery slope arguments. For example, in debates on euthanasia, it has been argued that allowing euthanasia would increase the termination of life without explicit request of the patient, would increase pressure to ask for euthanasia or increase ending of life in vulnerable groups (for example with less serious illness). Most of these arguments are in fact issues that could be studied empirically. Therefore, research can illuminate whether such arguments actually occur.

However, there is a limit to the role of empirical data. As Brody notes: "When the moral issues are fundamentally non-consequentialist, empirical data will largely be irrelevant and may even confuse the discussion by drawing attention to the wrong issues" (1993). The study of consequences is a necessary but not sufficient condition for being ethical. Focusing extensively on consequences might completely ignore fundamental deontological considerations.

An interesting example is the case of predictive genetic testing in minors. In view of potential harms, professional guidelines have always advanced a cautionary position towards predictive genetic testing. This position has often been subject of criticisms because of the fact that these harms were more speculative and not based on actual evidence. (Duncan and Delatycki 2006) In particular, related to susceptibility testing for common disorders, there is also a renewed debate over the possibility that such testing might lead to negative psychosocial effects. Based on the absence of proof of harms, some have argued that the rationale not to accept genomic testing is exaggerated and have therefore argued that parents should be entitled to request testing. (Nordgren 2014) Nevertheless, focusing the debate on the acceptability of this type of testing due to the absence of harms to individuals undergoing genomic testing might distort the focus of the debate from other relevant issues. In particular, testing minors denies them the opportunity to make these decisions for themselves later on in life, and questions to what extent minors have a right not to know. (Borry et al. 2014).

9.4.4 Normative Weighing

In this step we advance a reciprocity between the empirical and normative approach, i.e. an exchange between "the moral meaning found in the empirical context, and the moral meaning found in the various norms contending for application in a concrete discussion" (Borry et al. 2004). As described earlier:

> this means that a certain suggested course of action or moral judgement should acquire the necessary moral legitimacy. What happens then is that moral justification is granted to a certain argumentation and conclusion, which amounts to presenting sufficient grounds to support it (Beauchamp and Childress 2001). In our view, moral legitimacy cannot be granted on the basis of an inductive, bottom-up perspective alone. Empirical research cannot determine what is good or evil, right or wrong, by itself. The inductive approach runs the risk of lacking a critical attitude and assigning a sacred meaning to the facts without testing them against normative principles and theories. Persuasive argumentation must not only be based on empirical data, it must also explain and critically evaluate the values or ethical theories that are being presupposed (Nilstun et al. 2000; Nilstun and Sjökvist 2001).

9.4.5 Evaluating the Effects of a Decision

Finally, it would be ethically unacceptable not to consider the consequences and effects of certain decisions. Empirical research can hereby describe the actual effects of a decision and provide relevant information that might effect a new position. As described earlier, "since moral truth cannot be fixed once and for all, moral choices must be subject to continual question and, in some cases, adapted. Bioethics is a never ending process of evaluation and re-evaluation. Once a decision has been made, it is the moral responsibility of the decision-makers to be sure that the decision has no unforeseen and unintended effects or side-effects. The critical function of theoretical ethics here is to make sure that attention to effects does not, in the end, result in a victory of consequentialism (if this is not the appropriate moral theory). Empirical researchers must map out the actual effects of a decision and provide data that can foster further ethical reflection." (Leget and Borry 2010) Moreover, Sulmasy and Sugarman suggest that it might be important to study "actual behavioural adherence" to a new norm (2001).

9.5 Conclusion

Empirical research in bioethics is an expanding field whose results fundamentally question their place in normative key assumptions, moral hypotheses and/or ethical theories. Thinking about the contribution of empirical approaches to bioethics necessarily walks the fine line of the naturalist fallacy. This leads to sceptics accusing ethicists who are too focused on empirical research of attempting to *derive* normative statements from everyday facts. If one sees oneself as such a researcher, one can reply to this concern that there are no hard facts to begin with, as all facts are always of an implicitly normative nature. It will then, however, become difficult to explain why a division between facts and values in one's research remains relevant to begin with. And the question follows, why certain facts were chosen over others in a particular study. Therefore, this path leads nowhere. Thus, it might be better to reply to the sceptics using Aristotle. One might say that Aristotle, who himself based his normative-ethical deliberations on the everyday world, had already attempted to denote *phronesis* as being a type of virtuous wisdom, which was *oriented* towards practical action and therefore, towards practical facts. *Orienting* something towards something else is hereby already a different procedure than 'deriving,' as 'orienting' means that an ethicist attempts to guide his moral questions alongside the real world. As in our narrative, snorkelling gives insights that might not have been gained otherwise. As Alexander Kon expressed:

> While many tenets of bioethics have been derived from philosophical inquiry, empirical research is necessary as well to move forward in a well-reasoned and well-informed direction. Clearly, philosophical inquiry and empirical studies can work in concert, with each informing the other. As with all aspects of bioethics, the input of different experts with different training and techniques often yields the best results. The bioethics community should embrace both forms of inquiry as valid approached to normative ethics. (2009)

Suggested Readings

Ives, J., M. Dunn, and A. Cribb, eds. 2016. *Empirical Bioethics: Theoretical and Practical Perspectives*, (Cambridge Bioethics and Law). Cambridge: Cambridge University Press. https://doi.org/10.1017/9781139939829

Combining discussions of meta-ethical challenges, examples of different methodologies for integrating empirical and normative research, and reflection on the challenges of conducting and publishing such work, this book is a nice introduction to the relationship between empirical and normative approaches in bioethics.

Widdershoven G., T. Hope, L. van der Scheer, and J. Mcmillan, eds. 2008. *Empirical Ethics in Psychiatry*. Oxford: Oxford University Press. ISBN: 9780199297368.

The book provides a nice collection on articles that discuss the possibility of empirical ethics from a theoretical standpoint and an overview of the history of empirical medical ethics in general. Moreover, the book provides examples of empirical research with ethical implications in the domain of psychiatry.

References

Ackerman, T.F. 1980. What Bioethics Should Be. *Journal of Medicine and Philosophy* 5: 260–275.

Beauchamp, T.L., and J.F. Childress. 2001. *Principles of Biomedical Ethics*. New York: Oxford University Press.

Borry, P., P. Schotsmans, and K. Dierickx. 2004. What Is the Role of Empirical Research in Bioethical Reflection and Decision-Making? An Ethical Analysis. *Medicine, Health Care and Philosophy* 7: 41–53.

———. 2005. The Birth of the Empirical Turn in Bioethics. *Bioethics* 19: 49–71.

Borry, P., M. Shabani, and H.C. Howard. 2014. Is There a Right Time to Know? The Right Not to Know and Genetic Testing in Children. *The Journal of Law, Medicine & Ethics* 42(1): 19–27.

Braddock, C.H., 3rd. 1994. The Role of Empirical Research in Medical Ethics: Asking Questions or Answering Them? *Journal of Clinical Ethics* 5: 144–147.

Brody, B.A. 1993. Assessing Empirical Research in Bioethics. *Theoretical Medicine and Bioethics* 14: 211–219.

de Vries, R., and P. Conrad. 1998. Why Bioethics Needs Sociology. In *Bioethics and Society. Constructing the Ethical Enterprise*, ed. R. de Vries and J. Subedi. Upper Saddle River: Prentice Hall.

Duncan, R.E., and M.B. Delatycki. 2006. Predictive Genetic Testing in Young People for Adultonset Conditions: Where is the Empirical Evidence? *Clinical Genetics* 69(1): 8–16.

Fox, R.C., and R. de Vries. 1998. Afterword: The Sociology of Bioethics. In *Bioethics and Society. Constructing the Ethical Enterprise*, ed. R. de Vries and J. Subedi. Upper Saddle River: Prentice Hall.

Fulford, W.K.M. 1994. Not More Medical Ethics. In *Medicine and Moral Reasoning*, ed. W.K.M. Fulford, G.R. Gillet, and J.M. Soskice. Cambridge: Cambridge University Press.

Kon, A.A. 2009. The Role of Empirical Research in Bioethics. *The American Journal of Bioethics* 9: 59–65.

Leget, C., and P. Borry. 2010. Empirical Ethics: The Case of Dignity in End-of-Life Decisions. *Ethical Perspectives* 17: 231–250.

Leget, C., P. Borry, and R. de Vries. 2009. 'Nobody Tosses a Dwarf!' The Relation Between the Empirical and the Normative Reexamined. *Bioethics* 23: 226–235.

Miller, F.G. 2002. Ethical Significance of Ethics-Related Empirical Research. *Journal of the National Cancer Institute* 94: 1821–1822.

Molewijk, B., A.M. Stiggelbout, W. Otten, H.M. Dupuis, and J. Kievit. 2004. Scientific Contribution. Empirical Data and Moral Theory. A Plea for Integrated Empirical Ethics. *Medicine, Health Care and Philosophy* 7: 55–69.

Nelson, J.L. 2000. Moral Teachings from Unexpected Quarters: Lessons for Bioethics from the Social Sciences and Managed Care. *Hastings Center Report* 30: 12–17.

Nelson, H.L. 2014. *Stories and their Limits: Narrative Approaches to Bioethics*. New York: Routledge.

Nilstun, T., and P. Sjökvist. 2001. From Fact to Recommendation: Explicit Value Premises Make the Conclusion more Convincing. *Journal of Internal Medicine* 249: 121–126.

Nilstun, T., G. Melltorp, and G. Hermerén. 2000. Surveys on Attitudes to Active Euthanasia and the Difficulty of Drawing Normative Conclusions. *Scandinavian Journal of Social Medicine* 28: 111–116.

Nordgren, A. 2014. Neither as Harmful as Feared by Critics nor as Empowering as Promised by Providers: Risk Information Offered Direct to Consumer by Personal Genomics Companies. *Journal of Community Genetics* 5(1): 59–68.

Porz, R. 2008. Zwischen Entscheidung und Entfremdung: Patientenperspektiven in der Gendiagnostik und Albert Camus' Konzepte zum Absurden; eine empirisch-ethische Interviewstudie, Mentis.

Sugarman, J. 2004. The Future of Empirical Research in Bioethics. *The Journal of Law, Medicine & Ethics* 32: 226–231.

Sulmasy, D.P., and J. Sugarman. 2001. The Many Methods of Medical Ethics (Or, Thirteen Ways of Looking at a Blackbird). In *Methods in Medical Ethics*, ed. D.P. Sulmasy and J. Sugarman. Washington, DC: Georgetown University Press.

van der Scheer, L., G. Widdershoven, G. van Thiel, and J. van Delden. 2004. Theory and Methodology of Empirical-Ethical Research. In *Engaging the World: The Use of Empirical Research in Bioethics and the Regulation of Biotechnology*, ed. S. Holm and M. Jonas. Amsterdam: IOS Press.

van Willigenburg, T., A. van den Beld, F. Heeger, and M. Verweij. 1993. *Ethiek in praktijk*. Assen: Van Gorcum.

Weisz, G. 1990. Introduction. In *Social Science Perspectives on Medical Ethics*, ed. G. Weisz. Dordrecht: Kluwer.

Chapter 10
Normativity and Global Bioethics

Henk ten Have

Abstract Human rights are currently weakly defended. Many doubt their effectiveness and politicians often prioritize economic and military considerations. This chapter examines the implications of this weakening of human rights discourse. Is global bioethics losing its normative force? This question raises the issue of normativity in global bioethics. It is argued that normativity in global bioethics differs from that in ethics in general since it is a specific area of ethics with specific values. This specific normativity is reflected in theory and practice. As a moral discourse, global bioethics is broader than bioethics which is usually focused on the individual perspective. Global bioethics also has a larger repertoire of practices, including for example advocacy. The normativity of global bioethics furthermore is dialectical; it moves between the global and local levels so that it is continuously developing and transforming.

Keywords Advocacy · Global bioethics · Human rights · NGOs · V.R. Potter · UNESCO · United Nations · Values

10.1 Human Rights

"I don't care about human rights." This was the response of the newly elected Filipino President Rodrigo Duterte to the accusation of the United Nations' High Commissioner for Human Rights in September 2016 that he was lacking understanding of human rights institutions and principles. The president, elected by a huge majority of the population had started a war against drugs that resulted in the extrajudicial killing of more than 2000 people within the first year of his presidency. Duterte argued that corrupt journalists, drug dealers, and gang members were legiti-

H. ten Have (✉)
Center for Healthcare Ethics, Duquesne University, Pittsburgh, PA, USA
e-mail: tenhaveh@duq.edu

J. Gielen (ed.), *Dealing with Bioethical Issues in a Globalized World*, Advancing Global Bioethics 14, https://doi.org/10.1007/978-3-030-30432-4_10

mate targets of assassination. Since illegal drug use is a rampant problem in the country, human rights, in his view, should not be used as an excuse to do nothing against it.[1] The High Commissioner raised the question: "Are human rights exclusively a national issue?" He argued that governments have the responsibility to uphold their human rights obligations and to respect the standards within their territory. But human rights are more; they are the rights of all people, in all countries, because they are human. Therefore, they are also the concern of all of us. "Human rights are universal, indivisible and interdependent; if States pick and choose which rights they will uphold, the entire structure is undermined," the Commissioner said.[2]

"Putin is a human rights abusing oligarch" is the heading of a recent article in *The Guardian*.[3] The president of the Russian Federation is blamed for arguing that the country's constitution is taking priority over international human rights law. Nongovernmental organizations are stigmatized as foreign agents; freedom of expression and association are increasingly restricted; violence and discrimination against LGBT people is widespread[4]; and political opponents and journalists are murdered. Since Putin's return to the presidency in 2012, the human rights climate in the country has seriously deteriorated. The Russian government does not regard human rights as universal principles that all human beings should enjoy, but as dimensions of a Western, liberal lifestyle. This rejection of human rights is connected to specific bio-politics that emphasize Russian "traditional values." This is affecting bioethics discourse within Russia within the fields of sexuality, adoption, fertility, homosexuality and family planning. It has led to legislation on public hygiene but also to banning adoption of Russian children by U.S. citizens (Makarychev and Medvedev 2015).

These two examples demonstrate that discussing human rights and global ethics today is problematic. It is not only the case that human rights do not seem a serious concern for heads of state and governments. At the same time, the examples show that the touted impact of globalization and the U.N. system in this aspect is rather limited. Moreover, traditional champions of human rights have taken a low-key approach. For instance, in the US, who traditionally prides itself on its human rights advocacy, the Obama administration down played human rights and freedoms in its foreign policy (Pierce 2016) while for the Trump administration human rights do not have priority at all. Another example is human trafficking, which is one of the fastest growing forms of human rights abuse. Governments are regularly accused of not doing enough against this new form of slavery.[5] Although there are many courageous human rights advocates in all regions of the world, there is a need for more

[1] http://www.aljazeera.com/news/2016/08/rodrigo-duterte-human-rights-160806211448623.html

[2] http://globalnation.inquirer.net/144831/un-hits-duterte-for-lack-of-understanding-of-human-rights-institutions#ixzz4OrLQaAzE

[3] Owen Jones: Putin is a human rights abusing oligarch. The British left must speak out. *The Guardian*, 26 January 2016.

[4] https://www.hrw.org/world-report/2016/country-chapters/russia

[5] UK Human Rights Blog: UK not doing enough to combat human trafficking and domestic slavery. November 28, 2012. https://ukhumanrightsblog.com/2012/11/28/uk-not-doing-enough-to-combat-human-trafficking-and-domestic-slavery/

human rights defenders at the global level such as Nelson Mandela, Malala Yousafzai, Andrei Sakharov, and Shirin Ebadi, because the United Nations system itself is affected by hypocrisy. The Human Rights Council is the UN body responsible for promoting and protecting human rights around the globe. However, many members of this body are notorious violators of human rights. Saudi Arabia, which has a poor record in human rights, managed to be re-elected to the Human Rights Council in October 2016. The Philippines has been a member since 2012.

Many politicians not only discredit and ridicule international human rights discourse, and, at best weakly defend it in practice, they, also, increasingly criticize it from a theoretical point of view as ideological and ineffective. Duterte's conduct is reminiscent of the debate about Asian values in the 1990s. Authoritarian leaders at the time, for example in Malaysia and China, argued that Asian people have their own values such as harmony and orderly social behavior that are substantially different from western values that, in the leaders' opinion, lead to immorality, violence and decadence (Ten Have 2016). The same message is repeated by Duterte and Putin: human rights are not universal but an instrument of western countries to impose their values on the rest of the world. Similar arguments are elaborated in discussions about global bioethics. As a new and broader concept, global bioethics not only includes planetary concerns but also encompasses environmental and social aspects of global health. Is it possible to develop and justify ethical principles that should apply across the globe?

Another case against human rights argues that international human right law is a great moral achievement with the noble intention of protecting the powerless and the vulnerable, but that in reality it is ineffective and has not improved human well-being. All major human rights treaties are ratified by more than 150 countries, but in many countries the rights articulated in these treaties are continually violated (e.g. non-discrimination of women; prohibition of child labor) (Posner 2014).

10.2 Implications for Global Bioethics

What is the impact of the current uncertainty and discrediting of human rights language on the question of the normativity of global bioethics? Because global bioethics is based on this language, and specifically the idea that human rights are universal, it is more vulnerable to skepticism than mainstream bioethics that is based on ethical principles justified by philosophical reasoning. Global bioethics is losing its normative force if human rights are losing their normative power.

In the field of ethics in general, moral discourse is regarded as essentially normative. Contrary to descriptive ethics and meta-ethics, moral discourse issues prescriptions (oughts) and directives. Normativity is precisely what makes ethics different from other (philosophical and scientific) approaches of human existence. Normative ethics proceeds from a moral point of view that universally applies to everyone and that intends to assess the world in terms of good and bad. This interpretation of ethics in general has always been contested by relativism. Similar criticism is continuously directed against global bioethics.

No one contests that global bioethics deals with important moral questions. Is it ethically justified that poor people in developing countries are selling their kidneys? Should women in Cambodia be allowed to outsource their wombs to couples in Australia? Is it ethically acceptable that 1.5 million people die each year from tuberculosis while this disease is almost completely curable? Should global trade be more important than global health? Global bioethics searches answers to this kind of questions. However, although critics agree that these questions should be answered, they argue that global bioethics cannot do that. In their view, global bioethics does not provide a framework of normative reasons that apply universally. In their opinion, global bioethics cannot provide a convincing justification of the importance of a particular ethical principle (e.g. justice or personal autonomy) over another in case there is a conflict between principles. Some scholars even argue that global bioethics is a mission impossible. It can never be normative since agreement on moral issues does not exist (Engelhardt 2006). These scholars believe that the only thing we can do is to describe and analyze the conditions in which such ethical questions arise, and search for solutions and answers in a pragmatic way. Instead of universality, – they argue – the discourse of global bioethics refers to diversity. Studies on global health issues, therefore, celebrate the enormous heterogeneity of religious, cultural and social value systems instead of looking for uniformity.

Like the current criticism of human rights, critics of global bioethics have argued that it is not clear what values and principles are really global, i.e. independent from an already existing moral point of view. If the global nature of values cannot be proven, so-called global values remain always biased. It is also argued, more fundamentally, that global or cosmopolitan ethics not only does not exist in reality but principally cannot exist. Global bioethics states that a global framework of universal principles does exist, notably the ethical principles of the Universal Declaration on Bioethics and Human Rights, adopted by UNESCO in 2005. Critics point out that any framework of universal principles is a fiction. It is the outcome of political consensus and practical agreement. It cannot be justified by rational argument, and, thus lacks normative force. Such a point of view echoes the current denigration of human rights. Global bioethics would, then, be a noble idiom without impact, a new form of colonialism, or Western ethics in global clothes (Ten Have 2016).

Without going deeper into this debate and exploring what normativity is and how it can be justified, I would like to go in another direction and ask whether there is a difference in normativity between ethics in general and global bioethics specifically.

10.3 Normativity in Global Bioethics

Compared to ethics in general, the issue of normativity within global bioethics is in a certain way pre-formatted. Because of the specific characteristics of this area of ethics, normativity is structured and directed in specific ways.

Global bioethics is first of all concerned with ethical queries related to health and disease, medicine, healthcare and medical technologies. This is the focus on 'bio.' It means that many ethical subjects are outside of its scope (such as violence, peace, and friendship). As bioethics, it presupposes the priority of a particular value context in which promotion of health and relief of suffering have more prominence than other values. This selective scope and aim produce already a normative context whenever a concern is taken into bioethical account. This can be illustrated by the fact that people involved in global bioethics consistently prefer the term female genital mutilation over female circumcision. By using the former term, the practice is not just labelled as a bioethical problem, it is, at the same time, interpreted as an ethical wrong (Ten Have 2016). Not only is the practice no longer framed as female circumcision, which does not have a normative connotation, but the phenomenon is now already inserted in a normative framework as mutilation. Within a framework with health as pre-given and overriding value, intentionally damaging the body cannot be justified. Compared to ethics in general, bioethics is operating within a field of particular concerns and problems that assumes from the start a specific normativity: everything that promotes health and relieves suffering is good; and, because health and disease are universal phenomena, everybody will agree with this normative goal. Nobody will argue that bioethics should promote human suffering. This implies that a normative position is intrinsic in the concept of bioethics. That normative position is balanced and critically considers the particularity of each health-context. For instance, if women sell their bodies for sex it is prostitution, which could be disapproved on moral grounds, but if a woman sells her body for the purpose of surrogate motherhood the "sale" can be justifiable if it is regulated and free from exploitation. Also, human trafficking is universally condemned. Governments as well as human rights groups do everything possible to eradicate it. But organ trafficking, often associated with it, is viewed somewhat differently in bioethics. From a bioethical point of view, it is argued that rather than eradicating organ trafficking, it would be more efficient to redefine it as trade and regulate it. While ethics in general condemns human trafficking as a violation of human dignity that should be completely banned, bioethics regards organ trafficking within a preset framework of health and survival as basic values, aiming to transform the phenomenon with the assumption that selling one's organs is not ethically problematic against the backdrop of these values.

The first indication showing that the notion of normativity is different in global bioethics, therefore, is that this specific area of ethics is operating with a specific set of values. Without assuming these values, bioethics would be either general ethics or a different type of applied ethics (e.g. engineering ethics or legal ethics). The second indication is that global bioethics is a different type of ethics: not individual but social ethics. In ethics in general, the primary moral question is: what should *I* do? In global bioethics on the other hand, the basic question is: what should *we* do? Global bioethics, in other words, is social ethics. This social-ethical nature of the global ethics discourse explains why the normativity associated with global bioethics differs from ethics in general and mainstream bioethics in particular. There are five reasons why global bioethics has that social-ethical nature.

First, global bioethical problems have a specific character. If a tuberculosis patient in Egypt cannot afford medical treatment, there is indeed a moral question: why is that patient not receiving treatment that can cure him or her, and how can this be justified? But the moral problem is beyond the level of individual cases. Global bioethics assesses the more general issue: why do people in developing countries not have access to affordable and effective drugs? The fundamental moral problem is generic and global. The global problem manifests itself at the level of individual patients, but it cannot be restricted in ethical discourse to individual cases. Normative analysis requires a broader perspective. Bangladesh, for example, faces a growing problem with fake drugs. Television station Al Jazeera interviewed a mother who lost her son due to lethal counterfeit medication.[6] The same day in that hospital, three other children died from the same drugs. These shocking individual cases illustrate a general phenomenon. WHO estimates that almost 10% of all drugs sold globally are fake, with 100,000 people dying each year as a consequence of the consumption of counterfeit drugs (Ossola 2015). Focusing on the case of an individual will not explain the ethical problem: it will be necessary to connect the individual case with the underlying problem at the global level. That analysis will also make clear that Bangladesh, with an inadequate infrastructure, a dysfunctional government, unreliable regulations, lack of enforcement, widespread poverty and corruption will never be able to solve the problem at its own. It will need the cooperation of other countries that are, however, less impacted by counterfeit medication.

Second, global bioethical problems are not addressable by individuals. These problems transcend the level of individual agency. Even if individuals could address them, it would be unfair to leave the problem to individual management since individuals have not caused these problems. Global problems cannot be addressed by individuals but ultimately require cooperation and solidarity.

Third, while the global bioethical principle of respect for diversity demands that people take into account diverging moral views, international cooperation and global governance aimed at addressing global challenges such as Ebola and Zika pandemics require agreement on basic ethical principles. The need for global governance, therefore, implies the search and development of common perspectives that provide a basis for practical actions. Global bioethics is not merely theoretical reflection. Due to the confrontation with serious challenges to health, global bioethics is forced to apply its ethical framework in practices and policies, even if this framework is shaky, uncertain, and provisionary. This need for governance also makes its normativity different from ethics in general.

Fourth, the social nature of global bioethics is articulated in its basic ethical principles. Being 'global' means that a bioethical problem is affecting in principle all human beings wherever they live. Of course, problems challenge individuals, who have to determine what they ought to do in response. Many people work as volunteers to provide aid to victims of disasters; they donate money to NGOs such as

[6] Al Jazeera: Bangladesh faces growing trade in fake drugs. http://video.aljazeera.com/channels/eng/videos/bangladesh-faces-growing-trade-in-fake-drugs/3638512284001;jsessionid=5A9DFF5D69311D7A01288F1062A23E93

Doctors beyond Borders, or work for them some time in difficult circumstances. Infertile couples are aware that commercial surrogate motherhood is available, and they have to decide if they want to use these services when they desire children. It is evident that global problems can be translated into individual ethical problems. But as typically global phenomena they do not first of all present ethical challenges at an individual level. One conclusion from the global nature of contemporary bioethics problems is that they force us to go beyond an individually focused ethics. The implication is that global bioethics needs to articulate ethical principles that go beyond the point of view of individual moral agents. This is also why global bioethics differs from mainstream bioethics. This difference can be illustrated with the experiences during the last few years with the Ebola epidemic in West Africa which have highlighted mismanagement and lack of solidarity in the international community (Ten Have 2016). The main concern of Western countries was to prevent the virus to spread, and protect their own vulnerability, not the vulnerable people in Africa. Nor did they have any concern for the complete lack of health infrastructure and health professionals to cope with the disaster in that part of the world. The debate especially in the mainstream bioethics community was about potential treatments and new vaccines, and under what conditions these interventions could be expedited. Bioethics simply chose to ignore the social, political, and economic context in which the pandemic emerged and expanded.

In order to meet such challenges, global bioethics needs a wider ethical framework, as advocated by Potter (Potter 1988). The final reference should be humanity, not the autonomous individual. More in particular, according to Potter, global bioethics should be focused on the goal of human survival. This goal can only be accomplished when there is a balanced social and ecological system for human life to flourish. Bioethics, therefore, should do more than presenting an ethical framework focused on the individual good: it should also refer to the common good as an anchor for social justice and environmental sustainability. That requires an ethical point of view that is shared by as many others as possible, and at least aspires to apply universally to all people across the world, assuming, however, that, in practice, full agreement will never exist. This is exactly what the Universal Declaration on Bioethics and Human Rights is doing (Ten Have and Jean 2009). It does not present a hierarchy among principles but a range of principles focusing on the individual human being itself (human dignity, benefit and harm, autonomy), other human beings (consent, privacy, equality), human communities (respect for cultural diversity), humankind as a whole (solidarity, social responsibility, sharing of benefits) and all living beings and their environment (protecting future generations and protection of the environment, the biosphere and biodiversity). This ordering means that the focus of ethical concern is not restricted to an individual point of view; ethical concern should equally take into account common social and ecological perspectives. Global bioethics, therefore, cannot be merely medical, social or environmental. It should include all perspectives at the same time. At a global level, bioethics should have a wider focus and be concerned with humanity as a whole. These specific moral concerns explain why global bioethics is associated with a specific normativity.

The fifth and last reason why global bioethics has a social-ethical nature has to do with the sources and roots of the global problems with which it is confronted. Contemporary global bioethical problems are not simply the result of globalization as such but of specific processes of globalization that are primarily driven by economic motivations (Ten Have 2016). In contemporary economics and politics, the world is first of all regarded as a global market. The 'market' has become the dominant metaphor for the organization of social life. In this neoliberal view, everything, from transportation and research to healthcare and education, should be transformed into commodities and services transacted in a market (Lafontaine 2014). Competition rather than cooperation is the core value. It is clear now that neoliberal policies have had a negative impact on global health (Davies 2010). Social inequalities have increased. In the US (2012), the top 1% claimed 23% of national income (twice as much as in 1980). At the same time, real wages for the majority of people have stayed put or fallen. Poverty has increased: the US has the fourth highest rate of poverty among OECD countries. Over 30% of all American children live in poverty. After Chile, it is the most unequal country in the world. The social fabric responsible for equality of opportunity is gone (Putnam 2015). The same situation has been created in Europe. Solidarity which used to be a principle of pride in European politics is dead. Growing inequality is now considered as a major risk for political stability, undermining democracy and political institutions (Schui 2014). The damaging effects of neoliberal policies are particularly clear in the breakdown of healthcare systems in many developing countries. Reduced expenditures for health and social services, privatization of care, lower salaries for healthcare workers, and introduction of user fees for patients have reduced access to health services for the majority of populations. In Latin America, health reforms have increased inequity and inefficiency. Entire populations are deprived from necessary treatment and medication, simply because the prices are unaffordable. Globally, each year, 2–3 million people die of tuberculosis. Effective treatment exists but 79% of tuberculosis patients do not have access to appropriate medication. Against this backdrop, global bioethics is redefined as a *social* ethics. If global problems are produced within a context of neoliberal policies and practices, it will be inevitable to critically address this context (Brody 2009). If global bioethics wants to address the global problems that human health is facing, it, therefore, needs a specific normativity.

10.4 Normativity in Theory and Practice

When global bioethics, as argued above, is characterized by a specific normativity, the question is how that will reflect in theory and practice. In comparison to mainstream bioethics, global bioethics has a broader moral discourse, and an enlarged repertoire of practices.

10.4.1 Broader Moral Discourse

Although global bioethics incorporates the value of personal autonomy and individual rights, it has introduced a moral discourse with a more extensive horizon (Ten Have 2016). First, it argues that human beings are not abstract and de-contextualized individuals: they are necessarily embedded in social structures. Beneficial social, cultural, economic and political conditions make flourishing in health possible. Second, global bioethics emphasizes that living together means that human beings share heritage and common goods. Common interests are not the aggregation of private interests. Individual persons as citizens are concerned with the common good; they are not merely rational consumers interested in maximizing benefits for themselves. Third, though individual actions and concerns are important, they will not bring about social transformation. Neoliberal globalization is not a fact of nature but the result of deliberate decisions of policy-makers, often without any involvement of populations. New forms of collective engagement and agency will be necessary to influence the systemic conditions producing global problems. This is precisely what the logic of neoliberalism tries to prevent by arguing that individuals have moral responsibility to care for health and the environment, and promoting that they make 'ethical' choices (organic and labelled food; generic drugs or 'ethical' clothing). In the neoliberal view, ethical choices are options offered by the market; the individual consumers have to decide whether or not they will buy them.

Critique of neoliberal globalization as the source of bioethical problems has re-activated a broader set of moral concepts such as human vulnerability, social responsibility, sharing of benefits, and future generations within bioethics discourse. These concepts will direct bioethical attention to structural determinants of health rather than individual decisions concerning care. Other concerns than those uniquely focussed on the individual can, therefore, become part of the bioethics agenda; moral issues can be framed in a different way; other solutions and approaches can be proposed.

10.4.2 Enlarged Repertoire of Practices

Global bioethics not only is a theoretical discourse but also includes a broader variety of practices than those of mainstream bioethics. One example is advocacy, which is not a part of more traditional bioethics discourse. Many NGOs, however, have been using advocacy as an instrument to press for changes not in rhetoric but in concrete practices, for instance in pressuring governments to translate the right to health into better access to medication (Ho 2013). Advocacy sometimes requires language and actions that go beyond what is traditionally considered acceptable in the academic discourse of mainstream bioethics. Jennifer Chan (2015, p. 177), in her book on AIDS activism, highlights the international trade system, and the role of WTO, as "the rot at the core of global governance today." Blaming economic injustice and structural violence as basic problems is unusual in bioethics, but it opens up new ways to analyze and redress global problems. Another example of the broader variety of practices in global bioethics is capacity-building. Its significance is recognized in mainstream bioethics but it is usually regarded as 'empowerment': a way to improve the capacities of individuals as responsible agents of change. The assumption is that improvements will only result from enhanced individual efforts. Global bioethics, on the other hand, argues that capacity-building should be based on promoting the common good, and will require collective action. Capacity-building is primarily a matter of global injustice and structural inequalities that require a '*moral* structural adjustment,' such as fair benefit sharing with populations and communities and, in this way, aiming at strengthening a healthcare system, which would benefit the entire populace.

10.5 Normativity as Dialectics

In global bioethics, a common view explaining how normativity works in practice is that it operates at two levels. On the one hand, there is a set of global principles on which traditions and cultures agree. This is expressed in international human rights language and elaborated into specific ethical principles. On the other hand, there are many efforts to articulate more specific bioethics standards within the context of particular religious and cultural settings. Global principles are always applied within local and specific cultures and traditions, and this necessitates continuous interpretation and articulation of global standards (Sullivan and Kymlicka 2007; Held 2010).

However, this two level view easily suggests a hierarchy, as if global principles are imposing a specific ethical framework upon local settings. It, therefore, reinforces the idea that globalization is imperialistic and colonizing. This interpretation of the practical functioning of normativity assumes that globalization and localization are two different processes. In reality, though, global and local intermingle and interact. The distinction between two 'levels' is therefore not appropriate. Global and local interact at the same level. Also, the universality of principles identified at the global 'level' is the outcome of interactions with the local 'level,' so that the global principles are in fact shaped by particular settings and approaches. These principles only come to life when applied in local settings; they will thus be transformed while local experiences will feed back into the global principles. It is, therefore, better to say that global bioethics has several constitutive components, rather than two levels. These components are involved in dialectic interaction. Global bioethics covers a global domain as well as local domains but also involves a process of continuous dialectic interaction with mutual exchange, inspiration, aspiration, articulation, deliberation, learning, and negotiation. Global bioethics is characterized by the duality of global and local. It is global in the sense that it assumes a universal ethical framework (as the result of intercultural dialogue and consensus) and, at the same time, local in the sense that this framework necessarily has to be applied in diverse multicultural settings. It is, furthermore, characterized by labor. It is not simply academic reflection but work in practice.

This dialectic interaction between global and local means that global bioethics is not imposed upon the rest of the world but nurtured and transformed by the world. Bioethics is not a product that can be imported and used regardless of the local culture, because it is itself part of culture in a process of exchange, which means adaptation and modification, and sometimes rejection. In this process of cultural exchange in many different settings, a consensus will gradually emerge on a comprehensive approach through scrutiny, analysis, debate, application, modification, and reinterpretation. It is exactly through such a process of exchange and negotiation that the Universal Declaration on Bioethics and Human Rights came about in the first place, as the beginning not the closure of a long process furthering the expansion of the circle of moral concern. The same process is at work during the never-ending implementation of the UDBHR. The lesson from current human rights scholarship is that applying global principles is 'globalization from below' rather than from above (Koh 1999; Risse et al. 2013; Hafner-Burton 2013). Rather than compliance with global principles, practical applications entail "domestication." Principles need to be transformed and internalized into domestic systems and local contexts. Usually, this is not done by governments but by non-governmental organizations and cooperating individuals. The challenge to translate the global principles into local settings and practices, therefore, is with the bioethics, scientific, and professional communities. It is precisely this work of domestication or vernacularization that typifies the normativity of global bioethics. Because there is a framework of global principles, it inspires local activity. It provides an emancipatory tool for vulnerable populations: women, racial minorities or indigenous people (Merry 2006). A very relevant example, here, is the right to health. Because the right to health was adopted by the United Nations, grass root movements could militate against their own governments for better access to medication. Some countries have, indeed, included this right in their Constitution, while, nonetheless, nothing has been done by the government to implement it in practice. But, the idea of the right to health could inspire movements of, for instance, HIV/AIDS patients to make the government act and provide adequate access to HIV treatment where this is unavailable (Ten Have 2016). Normativity means that transnational ideas and universal principles travel from the global arena down and from the local arena up. It is not compliance or enforcement. These ideas and principles do not compel but inspire. The current criticism of human rights therefore misses the point (Simmons 2015). Arguing that human rights are useless because states and intergovernmental institutions are no longer interested in them or active within the area of human rights does not recognize where the real action is: it is with domestic actors and stakeholders.

10.6 Conclusion

Global bioethics, therefore, is not a ready-made product but a process. It is the aspiration to realize the universal in the local. But, it is, first of all, a social ethics that goes beyond the view that ethics is primarily a matter of personal commitment and individual lifestyle. Global bioethics presents a horizon of reflection, analysis, and action that brings ethical principles associated with cooperation, future generations, justice, protection of the environment, solidarity, social responsibility, and vulnerability (back) into the debate of globalization.

Suggested Readings

Akhavan, Payam. 2017. *In Search of a Better World. A Human Rights Odyssey*. Canada: Anansi.

Human rights law does not exist for the powerful. It is aimed to protect the vulnerable, the marginalized and disposed. Akhavan shows how human rights are endangered the world over but also how they can provide hope.

Sikkink, Kathryn. 2017. *Evidence for Hope. Making Human Rights Work in the 21st Century*. Princeton/Oxford: Princeton University Press.

This book argues that human rights has worked, does work, and can continue to do so. Change is slow and the result of struggle, but in the long term, human rights movements have been vastly effective.

Ten Have, Henk. 2016. *Global Bioethics. An Introduction*. London: Routledge.

This is an introduction into the new and emerging field of global bioethics. It discusses theoretical background and practical applications. It argues that a broader conception of bioethics is needed, encompassing not only social and economic dimensions of contemporary healthcare but also environmental concerns.

References

Brody, Howard. 2009. *Future of Bioethics*. New York/London: Oxford University Press.
Chan, Jennifer. 2015. *Politics in the Corridor of Dying. AIDS Activism and Global Health Governance*. Baltimore: Johns Hopkins University Press.
Davies, Sara E. 2010. *Global Politics of Health*. Cambridge/Malden: Polity Press.
Engelhardt, Tristram H. 2006. *Global Bioethics: The Collapse of Consensus*. Salem: M&M Scrivener Press.
Hafner-Burton, Emilie M. 2013. *Making Human Rights a Reality*. Princeton/New York: Princeton University Press.
Held, David. 2010. *Cosmopolitanism. Ideals and Realities*. Cambridge/Malden: Polity Press.

Ho, Cynthia M. 2013. Beyond Patents. Global Challenges to Affordable Medicine. In *The Globalization of Health Care. Legal and Ethical Issues*, ed. I. Glenn Cohen, 302–317. Oxford/New York: Oxford University Press.
Koh, Harold. 1999. How Is International Human Rights Law Enforced? *Indiana Law Journal* 74 (4): 1397–1417.
Lafontaine, Celine. 2014. *Le corps-marche. La marchandisation de la vie humaine à l'ère de la bioéconomie*. Paris: Seuil.
Makarychev, Andrey, and Sergei Medvedev. 2015. Biopolitics and Power in Putin's Russia. *Problems of Post-Communism* 62 (1): 45–54.
Merry, Sally Engle. 2006. Transnational Human Rights and Local Activism: Mapping the Middle. *American Anthropologist* 108 (1): 38–51.
Ossola, Alexandra. 2015, September 17. The Fake Drug Industry Is Exploding, and We Can't Do Anything About It. *Newsweek*. http://www.newsweek.com/2015/09/25/fake-drug-industry-exploding-and-we-cant-do-anything-about-it-373088.html.
Pierce, Anne R. 2016. *A Perilous Path: The Misguided Foreign Policy of Barack Obama, Hillary Clinton & John Kerry*. Post Hill Press.
Posner Eric, A. 2014. *The Twilight of Human Rights Law*. Oxford: Oxford University Press.
Potter, Van Rensselaer. 1988. *Global Bioethics. Building on the Leopold Legacy*. East Lansing: Michigan State University Press.
Putnam, Robert D. 2015. *Our Kids: The American Dream in Crisis*. New York: Simon & Schuster.
Risse, Thomas, Stephen C. Ropp, and Kathryn Sikkink, eds. 2013. *The Persistent Power of Human Rights. From Commitment to Compliance*. Cambridge: Cambridge University Press.
Schui, Florian. 2014. *Austerity; The Great Failure*. New Haven: Yale University Press.
Simmons, Beth A. 2015. What Is Right with Human Rights. *Democracy. A Journal of Ideas* 35. http://democracyjournal.org/magazine/35/whats-right-with-human-rights/.
Sullivan, William M., and Will Kymlicka, eds. 2007. *The Globalization of Ethics*. New York: Cambridge University Press.
Ten Have, Henk. 2016. *Global Bioethics. An Introduction*. London: Routledge.
Ten Have, Henk, and M.S. Jean, eds. 2009. *The UNESCO Universal Declaration on Bioethics and Human Rights. Background, Principles and Application*. Paris: UNESCO Publishing.

Chapter 11
Bioethics in Secular, Pluralistic Society

Joris Gielen

Abstract The existence of a plurality of ethical perspectives based on the presence of diverse religions and worldviews in contemporary society can make it hard to reach consensus on ethical issues in healthcare. Not only do religion and worldview influence attitudes, practices, and policies in healthcare. Racial, ethnic, cultural, and religious stereotypes and biases based on the assumption that people of a different social group are hard or even impossible to comprehend may obstruct dialogue and lead to inequalities and disparities in healthcare. In this context, moral relativism, which claims that, all moral viewpoints being equal, they can only be assessed from their own particular perspective, does not offer solutions and only reconfirms the impossibility of moral dialogue. The idea of common morality offers a way out of this deadlock by providing a common language on the basis of which normative ethical discussion in healthcare becomes possible. Differences may remain at the level of particular moralities. The common morality creates an atmosphere in which empathetic understanding of other ethical perspectives is enabled. In the case of ethical conflict, this offers an opportunity to search for creative solutions that are acceptable to all parties involved.

Keywords Pluralism · Religion · Worldview · Bias · Religious stereotype · Moral relativism · Moral dialogue · Common morality

11.1 Introduction

Ethicists publishing in academic journals sometimes fiercely disagree about bioethical positions and present their view in elaborate arguments whose primary audience often is fellow academic ethicists. The academic nature of these debates in

J. Gielen (✉)
Center for Healthcare Ethics, Duquesne University, Pittsburgh, PA, USA
e-mail: gielenj@duq.edu

J. Gielen (ed.), *Dealing with Bioethical Issues in a Globalized World*,
Advancing Global Bioethics 14, https://doi.org/10.1007/978-3-030-30432-4_11

journals and at conferences should, however, not make us to lose sight of the fact that their significance resides in their potential impact on government policies, biomedical research and clinical practice. At the same time, the dissonance and complexity of ethical positions in academic journals could make us wonder whether real agreement will often not elude clinicians and researchers when they encounter real ethical issues in their practice. If ethics specialists find it so hard to agree on the right normative position on ethical issues, this may be equally hard or sometimes even harder for non-specialists such as most biomedical researchers, physicians, and nurses, but also patients and their relatives. In complex ethical situations, they may require assistance from qualified ethicists who are specifically trained not just to perceive the ethical intricacies of clinical cases, but, also, find ways towards solutions.

In order to deal with ethical issues that arise within the context of biomedical research and clinical practice, dedicated ethics processes have emerged. For research, there are the institutional review boards (IRB's), and, for the clinical context, there are ethics consultations and ethics committees. Yet, the fact that these processes exist, does not by itself explain how they are able to function within a multicultural society and why their recommendations are effective in contemporary healthcare and biomedical research where the people involved may adhere to substantially divergent value systems. Obviously, one of the reasons why researchers follow demands for changes in research protocols by IRBs and why clinicians, patients, and their relatives heed to recommendations by ethics consultants is that the latter have a certain level of authority, as one of the next chapters will argue. In the case of IRBs that authority may even be backed up by law if the law of the country requires approval of the research protocol by an IRB before any study can be started. However, that barely explains how IRBs and ethics consultations most often prevent or resolve conflict in a way that is satisfactory to all parties involved.

This chapter argues that the effectiveness of IRBs and ethics consultations resides in the fact that they function within a normative ethical framework that is shared by the participants. Application of that framework may not necessarily lead all parties involved in an ethical issue to reach the same conclusion, but it will, at least, enable them to communicate and explain the rationality of their conclusions to the other parties. To make that point, I will, first, analyze the ethical diversity that exists in the healthcare setting and describe the adverse effects of a perceived absence of a shared normative framework on health outcomes. Next, I will argue that relativism, i.e. the absence of a shared normative framework, is not a necessity. Common morality may provide a solution. Though, as we will see in the last section, that solution to overcome relativism is not necessarily straightforward, but dialogical and incorporates plurality and dissent.

11.2 Ethical Diversity in the Healthcare Setting

Several studies have shown that religion and worldview influence attitudes, views, and practices in healthcare (Baeke et al. 2011b; Broeckaert et al. 2009; Gielen et al. 2009). Until a few decades ago, in Western countries, ethical attitudes in healthcare were largely shaped by Christianity, which was the dominant religion in these countries as well as a driving force in healthcare. In Mediaeval times, Christianity had also spurred the origination of early hospitals, or hospices. Hospices had been places where members of Christian religious congregations cared for travelers and the ill as part of their religious vocation. Over the centuries, in Western countries, the link between Christianity and healthcare persisted. This becomes particularly clear when we consider that Florence Nightingale (1820–1910) and others who set out the first benchmarks of modern nursing in the nineteenth century did not entirely sever the ties between the care that nurses provided and Christianity. Although Florence Nightingale contributed to the transformation of nursing into a lay profession, her Christian religious inspiration was never far away (Bradshaw 1996). In the second half of the twentieth century, at least in the vision of the founders of the modern hospice movement, the ties between nursing care and religion were renewed. Cicely Saunders (1918–2005), who was a nurse herself before becoming a physician and who is generally seen as one of the founders of the modern hospice movement, made religion an integral part of palliative care. Cicely Saunders was convinced that religion was not only important to the dying patients and their family members, but also to the members of the palliative care team. She even pondered whether ideally professional palliative care givers should not be members of a specifically dedicated religious congregation with its own rule and discipline (Clark 2001). Expressing this kind of idea was not particularly problematic in the UK in the early 1960s. Christianity was by far the dominant worldview in British society, and, for most people, deriving inspiration from Christianity, or attempting to spread a Christian message without proselytizing may not have seemed objectionable. At that time, Cicely Saunders could hardly have imagined how society would change in the coming decades and how the UK and the Western world as a whole would evolve into multireligious societies under the influence of migration, but also the conversion of Westerners to other religions. These societies are, at the same time, secular, in the sense that religion no longer openly dominates most day-to-day interactions. That does not mean that religion exerts no influence anymore on people's behavior and attitudes, as studies on the influence of religion on ethical attitudes in healthcare also show. However, the influence of religion often is, in a sense, more covert and only becomes clearly visible at times of conflict.

Such conflicts can occur in healthcare at the level of the individual healthcare professionals, but, also, at the organizational level. Mission-driven healthcare organizations often have a specific religious identity. Think, for instance, about Roman Catholic hospitals. The organization's mission will ideally transpire in everything the organization does, and will also be visible as such. For instance, a Roman Catholic hospital may commit itself to the service of underserved populations based on the belief that by serving them, the organization serves Christ. That specific religious identity can become problematic when it clashes with secular rules and regulations. Roman Catholic hospitals may, e.g., refuse to perform certain kinds of abortions even when they are legally permitted, or, in regions where voluntary (active) euthanasia or physician assisted suicide are permitted, Roman Catholic healthcare organizations may not want to perform these within their hospital compounds. Catholic hospitals may also be reluctant to comply with specific legislation targeted at large employers, such as the requirement in the American Affordable Care Act that organizations provide access to their employees to contraceptive methods through health insurance.

At the level of the patient-clinician encounter, differences in religious views can lead to disagreement about the right course of action when difficult decisions regarding treatment have to be made. For instance, under the influence of Islamic and Jewish scholars who have argued that curative or life-sustaining treatment ought not to be forgone (al-Mousawi et al. 1997; Baeke et al. 2011a; Sachedina 2005), some Muslims and Jews may hesitate to agree to withdrawal of such treatment in an incompetent relative, even if the treating healthcare professionals, who may be of a different religious orientation, consider continuation of the treatment futile and would rather withdraw it. In such a situation, clinicians may ponder whether there is a viable way out of this conflict, because, at first sight, this situation may seem like an unsurmountable clash of worldviews. That impression may be further exacerbated by stereotypical notions that clinicians may have regarding people with a religious orientation or cultural or ethnic background that differs from their own.

When stereotypes and bias enter the patient-clinician interaction, health outcomes may be affected. Due to the close connection of religion with race and ethnicity, biased attitudes to people belonging to religious minorities are tied with the broader issue of racial and ethnic disparities in healthcare. Studies have repeatedly shown evidence of racial and ethnic disparities in healthcare (Smedley et al. 2002, pp. 38–77). The reasons behind these disparities are diverse and complex. They have to do with factors that are largely external to the immediate healthcare context, such as economic differences (people belonging to ethnic and racial minorities often have jobs with lower salaries and less benefits, which affect living conditions and access to healthcare), and differences in educational level (racial and ethnic minorities have less access to good education, which impacts health literacy), as well as healthcare variables at the patient level (e.g. patients' preferences for particular treatments, and potential mistrust of the medical system), at the health system level (e.g. language barriers and problems related to availability and access to services), and at the level of the clinical encounter itself, where stereotypes and bias coupled with clinical uncertainty can lead to discrimination in care (Smedley et al. 2002, pp. 80–179). Medical decision-making is only rarely a straightforward process and often difficult choices have to be made. It is largely up to the discretion of the treating clinician to decide which options he or she is going to recommend or offer to the patient. In such a context, there is a potential for unconscious bias or even discrimination. Due to the large number of factors that play a role in medical decision-making, it is hard to prove that discrimination on the basis of race and ethnicity is a reality in healthcare in Western countries. What we do know, however, is that when there is racial or ethnic discordance between the patient and the clinician, i.e. the patient and clinician belong to a different race or ethnicity, communication tends to be harder.

These communication difficulties are often rooted in the fact that people belonging to different races or ethnicities may speak different languages, or different dialects of the same language. However, the difficulties may, also, follow from the unarticulated assumption that people from a different racial or ethnic background cannot be understood because their cultural framework is irreconcilably different. That cultural framework may include particular styles of expression that are not easily understood by people from other cultures, but may also include beliefs about reality that guide opinions in healthcare. Here, we could think, for instance, of the different religious views that were mentioned above.

The fact that religion need not be a unifying factor in healthcare, but can, also, be an obstacle to communication is obviously not an unexpected finding. Yet, on the surface, it contrasts with Émile Durkheim's renowned theory of religion. Émile Durkheim (1858–1917), who is generally seen as the founder of academic sociology, was of the opinion that religion has the power to form communities of people sharing the same moral standards. Religion creates – what he called – a "moral community." Durkheim believed that religion's power to unite through morality is so important that he gave this aspect a central place in his definition of religion. According to him, "a religion is a unified system of beliefs and practices relative to sacred things, that is to say, things set apart and forbidden—beliefs and practices which unite into one single moral community called a Church, all those who adhere to them" (Durkheim 1995, p. 44). Durkheim explained that religious beliefs delineate the characteristics of the sacred, from which prescriptions and restrictions for behavior follow. He was convinced that religion regulated human acts so efficiently that religion was essential to social cohesion, and society would not be able to survive without it. Obviously, Durkheim was not writing about the kind of multicultural and multireligious societies that we know today. From that perspective, it may not be surprising that nowadays religion appears as more conflictual than in the monoreligious societies that Durkheim studied. Yet, interestingly, Durkheim also argued that, gradually, secular society would take religion's place with its own rituals and customs that would unite society. The end of religion would, therefore, not be the end of the moral community. The notion of a moral community transcending the limitations of religions is an appealing idea if we want to understand how normative bioethics can be understood in contemporary multireligious and multicultural society. This society is much more complicated than Durkheim could have envisioned because secularism has not superseded religion. Rather, a multiplicity of religious perspectives continuous to flourish alongside secularism. In such a context, normative bioethics is possible if we assume that there is something like a global moral community. Otherwise, we will have to concede to moral relativism. The problem is that Durkheim's theory of religion or secular society as foundation of this moral community does not seem to be tenable within our current multicultural and multireligious society. Does that necessarily leave us with moral relativism?

11.3 Overcoming Relativism

Moral relativism seems very attractive when we look at contemporary multicultural and multireligious societies. In such societies, there are often conflicts that seem to point towards irreconcilable religious or cultural differences. Think, for instance, about a Muslim man who is living with his family in a Western city and insists on his wife being treated by a female gynecologist even when no female gynecologist is available. Cursory observations of this kind may lead to the conclusion that differences in moral standards between cultures and religions are unsurmountable. That conclusion may be further reinforced by findings from anthrophony, because anthropology registers diversity between cultures. If the intercultural differences in moral attitudes and values are seen as irreconcilable, two different but related and equally erroneous arguments can be developed. The first is the idea that one culture will be superior to another. The work of several nineteenth-century anthropologists led to that conclusion. These anthropologists perceived an evolution of cultures from savage to modern with Western civilization at the pinnacle. In such an evolutionary framework, the moral standards of the West were seen as superior, and non-Western people had to acquire Western moral standards if they ever wanted to reach the same level of civilization. The second idea, cultural relativism, can be seen as a reaction against this (Macklin 1999, pp. 30–31). In cultural relativism, no longer is one culture seen as superior to the others. At heart, the theory of cultural relativism is egalitarian because all cultures are considered equal. Based on that general equality of cultures, in cultural relativism, all particular moral values imbedded in each culture are equal, too. Consequently, people from other cultures can no longer objectively criticize moral values of another culture. In this way, cultural relativism functions as an effective antidote against Western moral imperialism. The problem is that this approach throws out the baby with the bathwater. If the moral norms of cultures can only be criticized from within, any action, however cruel or inhuman, could potentially be morally acceptable. Ruth Macklin gives the example of the "Nazi-argument": if there are no universal norms, then the acts of the Nazis can only be criticized by members of the Nazi party (Macklin 1999, pp. 4–5). This is hard to accept, but this is where cultural relativism logically leads to.

That problem can only be overcome if we accept that there is some kind of universality of norms. The conundrum that we have to face is whether or not it is possible to think of such universality while avoiding the pitfalls of both moral imperialism, in which one culture imposes its moral values upon another, and cultural relativism, in which intercultural dialogue is seen as futile because moral conceptions operate only within culture-specific frameworks. In order to find an answer to that issue, we should have a second look at anthropological data. In the pursuit of the universality of norms, anthropology may actually be an ally rather than an adversary. Anthropology not only teaches that there are differences between cultures. A closer inspection of anthropological data on ethical attitudes may show that besides differences there is, also, much commonality, but at a deeper, more fundamental level.

The example of autonomy illustrates this. In contemporary bioethics, autonomy is a very important category. It also figures within Beauchamp's and Childress' principles of biomedical ethics (Tom L. Beauchamp and Childress 2009). Their model has been developed in the West and, if we follow the relativist framework, its application including that of the principle of autonomy will be necessarily restricted to the Western world. That argument seems to hold if we consider that Western societies are often said to value autonomy and individual decision-making while non-Western societies emphasize relationships and authority. Yet, we should wonder whether this representation of Western and non-Western societies is not simplistic. Anthropological data reveal a much more balanced picture. These data show that both the Western and non-Western approaches to decision-making have been misrepresented and, because of that, the differences between both have been needlessly enlarged.

Western bioethical literature often lauds the ideal of the informed patient who decides about future treatment autonomously, albeit with assistance of the physician. It is often conveniently forgotten that, for most of Western medical history, physician paternalism was the norm and only in the last few decades it has become common practice among physicians to reveal a terminal diagnosis to patients (Macklin 1999, pp. 93–94; Sisk et al. 2016). It would be hard to consider autonomous decision-making in healthcare a Western tradition. Developing countries are now going through the same process of realizing the importance of revealing diagnoses and truth-telling. Moreover, even in the West, some patients are, in reality, still left in the dark about their diagnosis and, even though there is general agreement that patients should be informed the debate regarding what or how much they should be told has been far from settled (Sisk et al. 2016). This leads to situations in which healthcare providers or family members take over decision-making regarding treatment from patients who do have decision-making capacity.

On the other hand, it would be wrong to state that autonomy is a uniquely Western phenomenon. It is true that in non-Western contexts many persons other than the patient play a decisive role in the decision-making process. For instance, in India, family members of patients often withhold medical information to a patient with advanced disease because the family is afraid that knowledge about diagnosis and prognosis will destroy the patient's hope and will, in this way, accelerate the disease progress or complicate emotional wellbeing (Chattopadhyay and Simon 2008; Raj 2012). Without understanding of what is going on and even being unaware that important decisions regarding treatment have to be taken, patients in India are routinely left out of the decision-making process. Often, family members make the decisions, but it also happens that the treating physician decides (Kalantri 2000).

Strange as it may sound, it would be wrong to conclude from there that autonomy is not an important value in Indian healthcare. The fact that patients' wishes are often not considered does not necessarily mean that this is how decision-making ought to occur according to all parties involved. Many patients in India are actually not satisfied with their limited opportunities to participate in decision-making regarding their treatment and would want more information regarding their disease (Raj 2012; Raja 2007). Out of 100 patients who were admitted to the department of gastrointestinal surgery at a tertiary hospital in New Delhi, 98 wanted information regarding diagnosis and treatment options (Sanwal et al. 1996). These patients do not just want to know what is going on with them, to some extent, they also want to be involved in decision-making. Although the patients may not be getting a chance to actually take decisions the ideal of autonomy is clearly present in their view on how the decision-making process ought to be. To be true, Indian patients may be more welcoming towards the involvement of family members and physicians in this process. The process may be more one of shared decision-making and shared autonomy, but, in essence, many Indian patients, too, want autonomy just as their Western counterparts. It only is a different kind of autonomy, in which the involvement of family and care providers is more outspoken.

In this example, autonomy appears as a universal principle, yet, not in an absolute way. Absolutism would imply that the principle is forever fixed without the possibility of exceptions or contextual flexibility. Universality, on the other hand, allows for contextual reinterpretation of the general principle. Universality does not exclude the possibility that many people will disagree with the principle or that those who do agree will fail to display behavior that is accordance with the principle (Macklin 1999, pp. 51, 177). Likewise, in India, autonomy is important for many patients, but in concrete situations less so for their family members or care-providers, who may, however, still assent to the principle of autonomy in a general abstract way.

If it is accepted, on the basis of this example, that autonomy may very well be a universal principle, it may also make sense to assume that there are other universal principles, as well. There may, then, be a set of universal principles that together constitute a universal or common morality. Within the field of bioethics, the most influential expression of the idea of common morality can be found in Tom L. Beauchamp's and James F. Childress' *Principles of Biomedical Ethics.*' From the 1989 third edition of their work onwards, Beauchamp and Childress have been gradually developing their view on common morality as the basis of their four core principles: respect for autonomy, nonmaleficence, beneficence, and justice. Although Beauchamp's and Childress' use of common morality has been criticized (Arras 2009; Azambuja and Garrafa 2015; DeGrazia 2003) and is not the only expression of it within bioethics, it is worthwhile to study their analysis given the wide impact that their book has in its various editions. Moreover, even if one would disagree with the concrete way in which Beauchamp and Childress present common morality, their work could still be seen as a relevant example that points towards a general approach to understand how normative ethical discussion and dialogue is possible in secular and pluralistic societies. The idea of common morality shows how normativity can be conceptualized in secular, pluralistic society through the incorporation of plurality and dissent.

11.4 Incorporating Plurality and Dissent

In order to be able to understand how normativity can incorporate plurality and dissent, it is helpful to have a brief look into how Beauchamp and Childress present common morality. In their view, common morality refers to moral "core tenets […] that are not relative to cultures, groups, or individuals." (Tom L. Beauchamp and Childress 2009, p. 3) Consequently, these core tenets are applicable to all persons in all places. The tenets contain "standards of action (rules of obligation)," such as the prohibition to lie, steal or kill, "moral character traits or virtues," such as honesty, truthfulness, and gratitude, as well as their opposite negative character traits or vices, such as dishonesty or cruelty. Beauchamp and Childress argue that human rights are also supported by the common morality (pp. 3–4).

Beauchamp and Childress are aware that there may be persons who may not accept the validity of the common morality and have no qualms about acting against its tenets. Therefore, they contend that the validity of these tenets is accepted by "all persons committed to morality." (p. 3) There is a seeming circularity in this argument in that the validity of the common morality is based on its acceptance by those who are committed to morality, and morality is assessed by acceptance of the common morality. But, Beauchamp and Childress are indeed right that there will always be people who commit clearly immoral acts such as murder or live immoral lives, yet this need not imply the impossibility of morality as such. Here, we will not deal further with this particular issue. What interest us more is how Beauchamp and Childress allow plurality and dissent regarding the content of morality to operate in close association with the common morality.

The fact that Beauchamp and Childress allow scope for plurality and dissent does not mean that, in their thinking, common morality would not be the same for everyone. The common morality does apply to all persons, although it might change over time. Beauchamp and Childress think that such change is theoretically conceivable, but they find it unlikely to happen. It may also be possible that principles will be added to the four principles that Beauchamp and Childress present in *Principles of Biomedical Ethics* (pp. 412–415). However, settling the issue of what the common morality's core tenets are is just a part of the discussion. The more contentious issue is deciding about the concrete meaning, and implications of these tenets. The above mentioned example of autonomy illustrates this point. Autonomy can be a part of the common morality, even if people in the West and in India disagree about its applications and implications. People in India may be of the opinion that the desire to protect the patient from potential harm caused by knowing a terminal diagnosis overrides respect for autonomy, while people in the West will most often disagree with this. The fact that, in healthcare in India, autonomy is often not given priority, does not necessarily mean that autonomy is not respected. It may imply that exceptions are more often considered permissible. In Beauchamp and Childress' argument, disagreement plays out not within the common morality itself, which is one and the same for everyone, but at the level of particular morality. Particular morality represents the concrete outcome when the general tenets or principles of the common morality are applied in concrete contexts. Common morality represents general norms, while the applied norms in particular morality are concrete, specific, and, therefore, not universal as the norms in the common morality. Differences between particular moralities can be explained by different interpretation of the common principles (pp. 2–10).

Within this framework of common morality and particular moralities, discussion, dialogue, and compromise are primarily to be had at the level of particular moralities. In this way, the common morality is, we could say, the common language on the basis of which ethical discussion in healthcare can take place in a secular, pluralistic society. Because the common morality offers a common language, it will be more helpful than specific ethical theories, such as Utilitarianism or Deontology, to work towards agreement on practical issues in healthcare or develop healthcare policy (pp. 411–412). These ethical theories are not shared by everyone and, therefore, they cannot function as a basis for dialogue, while common morality is shared by everyone.

The above mentioned example on the involvement of patients in decision-making will illustrate how this idea of common morality enables normative ethical discussion in healthcare. Two principles are central in this discussion: respect for autonomy and beneficence combined with non-maleficence. The healthcare team that wants to involve the patient in the decision-making process in India would emphasize respect for autonomy. For them, it is important that the patient's wishes are respected. The family members, on the other hand, would emphasize beneficence and non-maleficence. They want to protect the patient from harm and benefit the patient's emotional well-being by hiding a diagnosis and prognosis that they consider potentially harmful. As such, these two perspectives are irreconcilable. However, since the principles respect for autonomy, beneficence, and non-maleficence are part of the common morality, there is scope for discussion. The particular moralities need to be reconciled.

The fact that the family's decision not to inform the patient is based on the common-morality principles of beneficence and non-maleficence, which are shared by the healthcare team, could enable the latter to develop an empathetic understanding of the family's approach. At the same time, since respect for autonomy is part of common morality, too, and many patients in India want to be involved in decision-making, there is a possibility to educate the family and explain the importance of involving the patient. While respecting traditional practices and beliefs, one can still educate patient and relatives about alternatives without condemning their preconceived notions and attitudes. One approach for the healthcare team would be to suggest to ask the patient whether or not he or she wants to know about the disease. The staff could argue that his approach would also be truly beneficial to the patient because it would relieve the patient of uncertainty. In this way, there is a possibility for an ethical consensus that is acceptable to both parties and respects both the emphasis on respect for autonomy of the healthcare providers and the family's desire to act beneficently without harming.

This situation would become slightly more complex if it were to occur with participants of diverse ethnic or cultural background, as could be the case, for instance, in a hospital in the US with a white American physician caring for a patient of Asian-Indian origin. In such a case, for a resolution to materialize more than common morality would be necessary. A profound understanding of intercultural interactions and the practical ability to empathetically deal with people of diverse ethnic and cultural backgrounds would be required. This we call cultural competence, which will be further explored in the next chapter. Cultural competence creates the environment in which a fruitful and effective normative dialogue in healthcare becomes possible through the common morality.

11.5 Conclusion

With all cultural and religious diversity that exists in contemporary healthcare, we can wonder whether dialogue about ethical issues in healthcare is really possible. At the same time, we know that understanding and proper communication is indispensable if we want to resolve ethical conflicts in healthcare to the satisfaction of all involved parties. Therefore, we have to find an answer to the question how fruitful normative ethical dialogue can happen if everyone has his or her own culturally and religiously determined set of preconceived values that may contradict. Common morality offers a way to overcome moral relativism in healthcare by offering a common language to discuss ethical issues. It enables an empathetic understanding of other perspectives in particular moralities so that striving for creative solutions that are acceptable to all involved parties becomes realistic.

Suggested Readings

The following book by Ruth Macklin provides a critical analysis of cultural relativism and possible responses to it.

Macklin, R. 1999. *Against Relativism: Cultural Diversity and the Search for Ethical Universals in Medicine*. New York: Oxford University Press.

In the field of bioethics, the common morality theory has been most influentially defended by Tom L. Beauchamp, James F. Childress, and Bernard Gert. In the current article, we analyzed the practical implications of common morality. Readers interested in its theoretical justification will find this in the following works.

Beauchamp, T.L., and Childress, J.F. 2009. *Principles of Biomedical Ethics*. New York/Oxford: Oxford University Press.

Gert, B., Culver, C.M., and Clouser, K.D. 2006. *Bioethics: A Systematic Approach*. New York: Oxford University Press.

References

al-Mousawi, M., T. Hamed, and H. al-Matouk. 1997. Views of Muslim Scholars on Organ Donation and Brain Death. *Transplant Proceedings* 29 (8): 3217.

Arras, J.D. 2009. The Hedgehog and the Borg: Common Morality in Bioethics. *Theoretical Medicine and Bioethics* 30 (1): 11–30.

Azambuja, L.E.O.d., and V. Garrafa. 2015. A teoria da moralidade comum na obra de Beauchamp e Childress. *Revista Bioética* 23: 634–644.

Baeke, G., J.P. Wils, and B. Broeckaert. 2011a. Orthodox Jewish Perspectives on Withholding and Withdrawing Life-Sustaining Treatment. *Nursing Ethics* 18 (6): 835–846.

———. 2011b. 'There Is a Time to Be Born and a Time to Die' (Ecclesiastes 3:2a): Jewish Perspectives on Euthanasia. *Journal of Religion and Health* 50 (4): 778–795.

Beauchamp, T.L., and J.F. Childress. 2009. *Principles of Biomedical Ethics*. New York/Oxford: Oxford University Press.

Bradshaw, A. 1996. The Spiritual Dimension of Hospice. The Secularization of an Ideal. *Social Science & Medicine* 43 (3): 409–419.

Broeckaert, B., J. Gielen, T. Van Iersel, and S. Van den Branden. 2009. Palliative Care Physicians' Religious/World View and Attitude Towards Euthanasia: A Quantitative Study Among Flemish Palliative Care Physicians. *Indian Journal of Palliative Care* 15 (1): 41–50.

Chattopadhyay, S., and A. Simon. 2008. East Meets West: Cross-Cultural Perspective in End-of-Life Decision Making from Indian and German Viewpoints. *Medicine, Health Care, and Philosophy* 11 (2): 165–174.

Clark, D. 2001. Religion, Medicine, and Community in the Early Origins of St. Christopher's Hospice. *Journal of Palliative Medicine* 4 (3): 353–360.

DeGrazia, D. 2003. Common Morality, Coherence, and the Principles of Biomedical Ethics. *Kennedy Institute of Ethics Journal* 13 (3): 219–230.

Durkheim, E. 1995. *The Elementary Forms of the Religious Life*. Trans. K.E. Fields. New York: Free Press.

Gielen, J., S. van den Branden, and B. Broeckaert. 2009. Religion and Nurses' Attitudes to Euthanasia and Physician Assisted Suicide. *Nursing Ethics* 16 (3): 303–318.

Kalantri, S.P. 2000. Informed Consent in Public Hospitals. *Issues in Medical Ethics* 8 (4): 116–117.

Macklin, R. 1999. *Against Relativism: Cultural Diversity and the Search for Ethical Universals in Medicine*. New York: Oxford University Press.

Raj, E.H. 2012. The Changing Scenario in Oncologist-Patient Communication: We Need to Adapt… …The Cherny Article Reviewed. *Oncology* 26 (1): 48–55.

Raja, K. 2007. Patients' Perspectives on Medical Information: Results of an Informal Survey. *Indian Journal of Medical Ethics* 4 (1): 16–17.

Sachedina, A. 2005. End-of-Life. The Islamic View. *Lancet* 366: 774–779.

Sanwal, A.K., S. Kumar, P. Sahni, and S. Nundy. 1996. Informed Consent in Indian Patients. *Journal of the Royal Society of Medicine* 89 (4): 196–198.

Sisk, B., R. Frankel, E. Kodish, and J. Harry Isaacson. 2016. The Truth About Truth-Telling in American Medicine: A Brief History. *The Permanente Journal* 20 (3): 74–77.

Smedley, B.D., A.Y. Stith, and A.R. Nelson, eds. 2002. *Unequal Treatment: Confronting Racial and Ethnic Disparities in Health Care*. Washington, DC: The National Academies Press.

Chapter 12
Narrative Authority: A Key to Culturally Competent Healthcare

Fahmida Hossain

Abstract Medicine is in the throes of disruptive change as it responds to the forces of globalization, multiculturalism, and dizzying technological progress. However, the practice of medicine and the design of healthcare remain problem-solving-oriented and mired in patriarchal, prescriptive responses to illness. In doing so medicine remains very directive, physician-oriented, and often blind to the various needs, beliefs, and understandings of the cultural and religious perspectives that patients bring to the examining room. In this changing landscape, contemporary healthcare needs revised practices and procedures to become narrative driven, i.e., founded on soliciting patients' personal narratives, and guided by clear organizational narratives. In turn, the organization and the providers must become culturally competent and globally responsive, ethical, and attuned to the humanness of the other.

If healthcare is to succeed in this global landscape, it must also come to see its role as healer in-service-to individuals and society at large. Technology, science, footprint expansions, and efficiency are means, not ends in themselves. Healthcare organizations should be redesigned in ways that encourage physicians to see patients as individuals, as others-as-self; patients should be taught to see and experience physicians as caregivers and healers, not as distant, mechanical experts. Healthcare must provide humanistic, narrative-driven connections that open a space for dialogue, consensus building, and healer-person partnerships. These notions and goals are aspirational. They constitute an on-going, attentive process to attain excellence, welcome the other, and create culturally attuned and responsive healthcare systems.

The chapter offers an approach called Narrative Authority, which serves as a foundational, organizational guiding principle for confronting the challenges contemporary healthcare faces. Narrative Authority is an approach that can positively transform a healthcare system by shepherding it to become other-oriented, dialogue-based, culturally sensitive, mindful of value judgments, and always story-dependent. Narrative Authority reflects the dialogic engagement of the provider and patient in partnership, together in the quest for effective patient care.

F. Hossain (✉)
Center for Healthcare Ethics, Duquesne University, Pittsburgh, PA, USA

J. Gielen (ed.), *Dealing with Bioethical Issues in a Globalized World*,
Advancing Global Bioethics 14, https://doi.org/10.1007/978-3-030-30432-4_12

Keywords Globalization · Cultural competence · Cultural responsiveness · Communication · Narrative authority · Organizational change · Healthcare · Storytelling · Personal narrative

12.1 Introduction

Exceptional contemporary healthcare requires a revision and the transformation of practices that have been based, almost exclusively, on privileged Western notions of individuality and autonomy. The time for this revision is now. Globalism is a sea change that has made the world 'flat'[1] and has brought countries and cultures into economic and, thus, health care exchange. Global healthcare delivery now requires practices and procedures that embrace the multicultural nuance patients carry with them to the consultation room. This means, cultural competency is a fundamental requirement for all involved in contemporary healthcare. The first important step along this path is to recognize, acknowledge, accept, and appreciate the distinctions different cultures offer as they enter the healthcare system. In this chapter, I contend that cultural competency tied to and guided by a complementary narrative-driven organizational culture are necessary for culturally competent, other-oriented perspectives and practices to arise. This revisioning can be best achieved by embracing and applying the concept of *Narrative Authority*.[2]

Narrative Authority serves as a foundational, organizational guiding principle. Narrative Authority is an approach that can positively transform a healthcare system by shepherding it to become other-oriented, dialogue-based, culturally sensitive, mindful of value judgments, and always story-dependent. Narrative Authority provides the perspective through which it becomes a matter of belief and practice to respect others, communicate effectively, listen actively, and respond reflectively. By practicing Narrative Authority one becomes other-oriented. This perspective eliminates the traditional distance between patient and physician and helps to co-create a partnership—a being-with—where both parties' voices are respected and heard. Narrative Authority is a means through which it is possible to become fluidly culturally competent in an increasingly

[1]The concept "flat world" is used by Thomas Friedman in his book *The World is Flat: A Brief History of the Twenty-first Century.* He argues that "the world is flat in the sense that the competitive playing field between industrial and emerging market countries is leveling; and that individual entrepreneurs as well as companies, both large and small, are becoming part of a large, complex, global supply chain extending across oceans, with competition spanning entire continents."

[2]The term Narrative Authority emerged during a series of February 2018 conversations with Frank Lehner. I am grateful for his insights, support, and mentorship through which I have developed my perspective on and applications for Narrative Authority.

multicultural world. Certainly, there is no complete or "final" level of cultural competency; instead, it is an aspiration, an on-going, attentive process to attain excellence, welcome the other, and create a culturally attuned and responsive healthcare system. This chapter further addresses how a healthcare system becomes—at its heart—culturally responsive, focused on building cultural competency, and dedicated to exercising Narrative Authority at all levels of the organization.

12.2 Recognizing and Accepting Cultural Differences

To be a successful and noteworthy healthcare system in the global marketplace, an organization must fully acknowledge and embrace the imperative to become culturally competent and culturally responsive. Globalization and multiculturalism have forced fundamental and, truly, paradigmatic changes to the entire breadth of healthcare practices. This is particularly evident in the changing shape of Western healthcare practices. This can be seen in emerging, new operating agendas and business models; the internationalization of research protocols and clinical trials; and revised organizational values, practices, and norms. Given the sea change that is occurring, healthcare must continue to recognize and overcome the White, Patriarchal, Christian, Western World biases that remain prevalent and retard necessary change. There has been positive change in this area, but they are not yet sufficient. Healthcare must become more sensitive and attuned to nuances and particularities of the many and varied cultures and diverse perspectives that increasingly intersect and make their way to the healthcare system. Healthcare can no longer operate effectively, efficiently, or ethically if it does not foundationally recognize and respond to the brilliant opportunities that cross-cultural intersections provide. Cultural competency is now a requisite for any reputable healthcare system. Thus, a revised orthodoxy—a re-visioned and shared narrative—is required that articulates and discloses the meaning for and reasons why culturally competency rests at the heart and spirit of healthcare itself. The current guiding narrative is deficient. It is mired in a misperception of humans as rational, isolated, autonomous decision makers, and organizations as calculative, efficient machines. Narrative Authority aims to reclaim the disappearing human element of healthcare as an antidote to cultural blindness and paternalistic, but still Western, biases. Narrative Authority helps to understand the actual illness from the patient's point of view. By understanding the narrative of illness, it becomes possible to provide medical care with humility, trustworthiness, and respect.[3]

---

[3] Rita Charon, *Narrative Medicine: Honoring the Stories of Illness*(Oxford: Oxford Univ. Press, 2008), 3–7.

12.2.1 Globalization and Its Impact

Globalization is an overpowering force tearing regionalism and the local and vernacular asunder. Globalism has a great homogenizing effect. The marketplace is controlled by a few large players who exert ever-greater influence. As a result, once inflexible national and cultural boundaries are dissolving. This gives rise to international mobility. People travel more easily, emigrate seeking better standards of living, or find themselves forced by natural and man-made disasters into safe havens in distant countries. As the world becomes more global, modern medicine is responding by creating a shared medical culture.[4]

As noted above, globalization blurs the distinctions between the local and global, and cultures mix as never before. The force of these changes necessitate that healthcare providers re-imagine the ethical landscape before them, not merely in terms of geographical distances, but rather as a world comprised as one human family.[5] Healthcare must become attuned to a variety of cultures, beliefs, perspectives that come for treatment and, in turn, develop and revise process and protocols to ensure that the "other's" perspective, i.e., her story, is taken into account. To this end, healthcare organizations must become more diverse, and reflect the changing nature of our ever-intersecting cultures and perspectives—the entanglement of stories.[6] Narrative Authority will help healthcare systems drive this change by defining illness from multiple points of view and consider culturally and socially determined beliefs on illness and disease.

As human we make sense of the world through narrative and storytelling. By nature we are integrated and immersed into a narrative universe where we find meaning and understanding; this helps us make sense of everyday practices. We are not isolated bodies gazing onto a fixed, immutable exterior world.[7] Who we are is inextricably woven into a relational web of meaning and practices. We are embodied—most often pre-reflectively—in a shared narrative world, a context full of common practices and norms. Now, with the intersection of cultures, different world-defining narratives are intersecting. This causes confusion, conflict, misunderstandings, anxiety, and often violence. This confusion is exacerbated by mistaking one narrative as privileged over other narratives. Doing so necessarily leads to forms of dehumanization and patriarchy. It places the other outside, as an observer and, in turn, eliminates her sense of identity and agency. For healthcare to flourish ethically in the global environment, leaders of healthcare organizations and policy makers must change their conceptual frameworks, which are often founded on

---

[4] Padela, Aasim I., Aisha Y. Malik, Farr Curlin, and Raymond De Vries. "[Re]considering Respect for Persons in a Globalizing World." *Developing World Bioethics* 15, no. 2 (2014): 99–101.

[5] "Why US Health Care Should Think Globally," *The AMA Journal of Ethics* 18, no. 7 (2016): 736–738.

[6] "Globalization of Healthcare," *Global Advances in Health and Medicine* 1, no. 2 (2012): 5–7.

[7] Rita Charon, *Narrative Medicine: Honoring the Stories of Illness* (Oxford: Oxford Univ. Press, 2008), 4–9.

presumed power and privilege. Decision makers must come to understand that all people present themselves as fully human, and this humanness is evident in and through their cultural stories. We must learn to solicit and "hear" those stories as shared, ethical access points where we "see" the other as self.[8] Healthcare organizations must change from the belief in the privileged universality of the predominant and powerful culture and embrace the validity and normativity of transnational cultures and diverse peoples. To do otherwise serves to violate international human rights. Healthcare organizations can lead the way in dissolving barriers, and set examples for overcoming rigid cultural, religious, and national stereotyping and prejudice. They can and must do so by creating cultures and practices that welcome patients and physicians from around the world—welcoming the other's story as a means to embracing her humanness, as means to provide ethical healthcare. As odd as this may sound, now is the time for healthcare systems to embrace the local—the story—within the global forces for expansion and standardization. In this era where diversity and multiculturalism are facts of life, stories are the means to building community and preserving agency and autonomy.[9] Embracing this reality lies at the heart of ethical healthcare.

Globalization creates business and wellness opportunities in healthcare through the exchange of medical products and shared services. However, complexities in financing, regulation, and the types of devices used in different countries often hinders the flow. Given the increased ease of travel and the enhanced speed of communication, policy makers should realize that it is time to make the medical exchange globally friendly and less complicated in terms of policies and procedures. The health system should focus on innovation, transformation, and collaboration that satisfy the global need as the world has essentially become a single place where different practices intersect and intermix.[10]

Healthcare systems that do not embrace this change will become isolated, less relevant, and will likely be driven from the marketplace. A global-serving healthcare organization that integrates healthcare ethics needs an integrated portfolio of product and services that will breakdown boundaries and tame complexity for people of all cultures. Besides breaking barriers and overcoming complexity of medical practices, the healthcare system that develops in this new environment must understand and address the moral and ethical implications of resource integration. There is a cultural sea change occurring: for a healthcare system to succeed it must now focus on the interdependent nature of shared medical practices and resource integration. This new age demands pluralism as a guiding principle.[11]

[8] Carlo Rovelli, Simon Carnell, and Erica Segre, *Seven Brief Lessons on Physics* (New York: Riverhead Books, 2016), 65–70.

[9] Leigh Turner, "From the Local to the Global: Bioethics and the Concept of Culture," *Journal of Medicine and Philosophy* 30, no. 3 (2005): 305–315.

[10] K. Lee, "For debate. The impact of globalization on public health: implications for the UK Faculty of Public Health Medicine," *Journal of Public Health* 22, no. 3 (2000): 253–255.

[11] Leigh Turner, "From the Local to the Global: Bioethics and the Concept of Culture," *Journal of Medicine and Philosophy* 30, no. 3 (2005): 306–312.

In addition to resource mobilization, healthcare has seen a significant expansion of knowledge sharing. Even though there remain significant knowledge and development divides in medicine between the East and West, globalization serves to bridge these gaps. For example, Cornell and Harvard Universities have developed satellite medical practices in the Middle East. Similarly, John Hopkins University has expanded its operations into Singapore and Abu Dhabi. Such initiatives are direct responses to new global knowledge-sharing opportunities. Despite the geographical distance, people can now receive the same or very similar high-quality healthcare services globally.[12] Knowledge-sharing and medical practices exchanges create a distributed healthcare system that must embrace and incorporate diversity and inclusion into its normative practices if it is to succeed.

"Medical tourism" is also a direct byproduct of globalization. Patients can travel abroad from their home countries to find less expensive and often better services. For instance, European countries attract American couples for fertility services, offering these services at a significant reduction of cost compared to the fees charged in the US. On the other hand, physicians from emerging countries migrate to developed countries to take advantage of higher wages and improved lifestyles.[13] As a result, there are increasing numbers of healthcare systems that are by necessity becoming more diverse. In fact, healthcare is not confined to medical practices only. Another impetus to change is that healthcare organizations compete with other systems within and outside the county. Insurance companies, NGOs, and political influence have significant impacts on the care delivery processes. These other groups are usually staffed with people from different cultural and ethnic backgrounds. Therefore, healthcare providers are further pressured to revise their practices—bridging cultural gaps—to offer the best possible care to their patients.[14]

It is predicted that by 2050, in the USA, the majority of the population will be people of color. Current "minorities and the people of color" will grow to 54% of the total US population. By 2050 the US population is expected to be 439 million of which 235.7 million will be "minorities." This is in good part an effect of globalization, and it clearly indicates how important it is for a healthcare system to become culturally competent and responsive. It is imperative that healthcare providers in the US become culturally sensitive if they are to properly address the large and diverse groups of people who will need and expect their services.[15] In this chapter the terms 'cultural competency' and 'cultural sensitivity' are used interchangeably.

Since health is a global public good and a basic human right, the healthcare system must, therefore, be culturally competent if it is to act in a culturally responsive

[12] Padela, Aasim I., Aisha Y. Malik, Farr Curlin, and Raymond De Vries. "[Re]considering Respect for Persons in a Globalizing World." *Developing World Bioethics* 15, no. 2 (2014): 98–100.

[13] C. Segouin, "Globalization in health care: is international standardization of quality a step toward outsourcing?" *International Journal for Quality in Health Care* 17, no. 4 (2005): 277–279.

[14] "Globalization of Healthcare," *Global Advances in Health and Medicine* 1, no. 2 (2012): 5–7.

[15] "Minorities expected to be the majority in 2050," CNN, accessed November 28, 2017, http://www.cnn.com/2008/US/08/13/census.minorities/index.html?_s=PM%3AUS

manner.[16] It is time to understand that the embrace of pluralism is necessary if a healthcare organization strives to provide ethical and excellent healthcare. The growing diversity within almost every society requires an embedded, interactive system where both theoretical knowledge and normative, other-oriented sense-making will prevail. The transformative power of Narrative Authority will serve any organization in its quest to become culturally competent and positioned to embrace the benefits of the future. Narrative Authority provides an approach that guides organizational culture change, and if followed, will, in turn, create normative practices through which individual providers act in culturally sensitive ways.

12.2.2 *Cultural Insensitivity*

The concept of culture refers to all the social practices and dominant values of any community. However, sometimes the recognition of culture invokes stereotyping. Cultural sensitivity occurs when one understands the nuances of a particular group, which, in turn, opens the possibility to be sensitive to and see the other, not through the gaze of an assumed stereotype, but as other, as human, and as self.[17]

A successful healthcare system must be culturally sensitive and responsive to social norms. Otherwise, the system gives rise to a form of cultural imperialism. A common criticism of a healthcare system is that it ignores the social and cultural aspects of the market that it enters. Most often in these cases healthcare systems are involved in bad and unethical practices that are driven by the quest for market dominance and profit. This is a form of neocolonialism. Neocolonialism occurs when developed countries use their resources and power to take advantage of poor and developing countries.[18] For instance, pharmaceutical companies conduct clinical trials in developing countries where there are fewer regulations and less stringent oversight.[19] In these instances, an alternative ethical framework should be created to hold developed countries accountable for abusing their power. Healthcare should heal and serve those in need. When market share is the key driver of the healthcare system, the practitioners may become less empathetic or corrupt, which gives rise to social and cultural discrimination, and ineffective "doctoring."[20]

---

[16] Tineke A. Abma et al., "Inter-Ethics: Towards An Interactive And Interdependent Bioethics," *Bioethics* 24, no. 5 (2010): 253–255.

[17] Leigh Turner, "From the Local to the Global: Bioethics and the Concept of Culture," *Journal of Medicine and Philosophy* 30, no. 3 (2005): 309–315.

[18] Karori Mbugua, "Respect for Cultural Diversity and the Empirical Turn in Bioethics: A Plea for Caution," *Journal of Medical Ethics and History of Medicine* 5 (2012): 2–5.

[19] Ruth Macklin, "Double Standards in Medical Research in Developing Countries," *SCRIPT-ed*, 2004, 639–640.

[20] Somnath Saha, Mary Catherine Beach, and Lisa A. Cooper, "Patient Centeredness, Cultural Competence and Healthcare Quality," *Journal of the National Medical Association* 100, no. 11 (2008): 1–9.

An example of how stereotypes can influence decision-making can be seen in diversified decision-making for white patients vs. nonwhite patients. For instance, nonwhite patients are generally prescribed less pain medication because of stereotyping; it is assumed *they* are likely to misuse the drugs.[21] Such experience instills, for minorities, a fundamental distrust in the healthcare system. As a result, the healthcare system fails in its obligation to provide equitable care.[22]

Ignoring or remaining insensitive to cultural differences indicates disregard for the new global normativity emerging in healthcare. Cultural insensitivity blinds leaders, administrators, and professional caregivers to the ethical necessity for change. Thus, emerging normative practices are discounted and the system acts as a status-quo social institution rather than a genuine, forward thinking—and practical—care provider.[23] Therefore, for a healthcare system to change, it should focus on knowing or soliciting the narratives of each culture, and commit to revising organizational practices to accommodate these different voices and perspectives.[24] New normative practices are often not being adopted in healthcare systems because of intolerance, bias, and bigotry that remain entrenched in multiethnic, pluralistic settings. Another limiting factor is that different practices of medical ethics, rooted in or coming from other multicultural communities, are merely discounted, ignored, or understood as deficient.[25] Therefore, a system and its stakeholders must find ways to become comfortable with and accept cross-cultural differences through an integrative narrative approach that opens the space to see others as guided by specific beliefs and perspectives. A culturally sensitive, narrative-driven healthcare system serves to improve the health outcomes of patients, increases the effectiveness of practitioners, promotes patient satisfaction, and makes an excellent healthcare system where ethical medical and business practices prevail.[26]

However, the notion of multiculturalism often hinders cultural sensitivity. Multiculturalism is often portrayed in a way that is negative. It misses the inclusive components and focuses on intercultural differences. In this way, people are characterized by group, race, and ethnicity. This way of looking at multiculturalism implies a form of blindness, which inhibits the ability to recognize *individuals* and, as a result, promotes stereotypical, paternalistic responses.[27] The healthcare system can

---

[21] Diana Jill Burgess et al., "Patient Race and Physicians' Decisions to Prescribe Opioids for Chronic Low Back Pain," *Social Science & Medicine* 67, no. 11 (2008): 1852.

[22] Somnath Saha, Mary Catherine Beach, and Lisa A. Cooper, "Patient Centeredness, Cultural Competence and Healthcare Quality," *Journal of the National Medical Association* 100, no. 11 (2008): 5–8.

[23] S. A. Buetow, "To Care Is to Coprovide," *The Annals of Family Medicine* 3, no. 6 (2005): 553–555.

[24] Clive Baldwin, "Narrative, Ethics and People with Severe Mental Illness," *Australian & New Zealand Journal of Psychiatry* 39, no. 11–12 (2005): 1023–1026.

[25] Leigh Turner, "Bioethics in a Multicultural World: Medicine and Morality in Pluralistic Settings," *Health Care Analysis* 11, no. 2 (2003): 114.

[26] Laurie M. Anderson et al., "Culturally competent healthcare systems," *American Journal of Preventive Medicine* 24, no. 3 (2003): 68–71.

[27] Leigh Turner, "From the Local to the Global: Bioethics and the Concept of Culture," *Journal of Medicine and Philosophy* 30, no. 3 (2005): 305–309.

counter this phenomenon and promote cultural sensitivity by tempering the notion of group identities. The system must develop practices based on the belief that each person has a unique cultural component, and also has a complementary individual identity that deserves and demands respect.[28] Narrative Authority offers a means to reveal each individual's distinctive perspective by engaging in and requiring a story-based, dialogue-oriented form of communication.

Each human is a narrative being. Humans understand and respond to the world by the continuity- and sense-making that our shared narrative provides. The shared narrative is the story that gives the everyday its meaning. It pre-reflectively provides the understanding of how things matter. The shared narrative makes thing intelligible and opens a space, and, within this space, the story conveys values and norms. Empathy and obligation are valued through the story's tenets.[29] The shared narrative—the common understanding of the everyday—helps to make things and practices intelligible, giving them significance. A shared narrative configures the tools, practices, and values of a society or community together into an inseparable 'weave of references'[30] and meaning.[31] While immersed within this larger narrative, we each find a way to embrace our imaginations of whom we want to be and how we want to live, given the possibilities provided by that larger narrative.[32] Regarding healthcare and wellness, we now find overlapping narratives and within these narratives, we each find and interpret the definition of illness and well-being. Thus a practitioner's understanding of illness may well be entirely different than the patient's. This is likely when the patient and the doctor come from different cultures. Given the growing intercultural makeup of healthcare, the understanding of medical practices by the practitioner and the patient are more complicated than ever before. This, again, is why the effects of medical technology, the medical model, and its accompanying practices must be reviewed, revised, and understood through a narrative that provides the perspective an authority to welcome and embrace the ethical nature of cultural sensitivity. Narrative Authority helps the practitioner understand the plight of the patient's complex story in this multicultural society by the power of narrative clinical imagination.[33]

[28] Padela, Aasim I., Aisha Y. Malik, Farr Curlin, and Raymond De Vries. "[Re]considering Respect for Persons in a Globalizing World." *Developing World Bioethics* 15, no. 2 (2014): 98–100.

[29] Rita Charon, *Narrative Medicine: Honoring the Stories of Illness* (Oxford: Oxford Univ. Press, 2008), 8–13.

[30] *Weave of References* is a term that that refers to how we generally find things meaningful or significant. It is the condition that makes it possible for any phenomenon to be the kind of thing it is. It is the intersecting, interconnected, inter-reliant set of references that configure human practices and concerns. The meaning of one thing is related to the meaning of all other things.

[31] Frank Lehner, *The Ingenious Workplace: An Antidote To Common Workplace Maladies* (Pittsburgh, PA: Naridus, LLC, 2012): 9–13.

[32] Clive Baldwin, "Narrative, Ethics and People with Severe Mental Illness," *Australian & New Zealand Journal of Psychiatry* 39, no. 11–12 (2005): 1022–1026.

[33] Rita Charon, *Narrative Medicine: Honoring the Stories of Illness* (Oxford: Oxford Univ. Press, 2008), 3–9.

Therefore, the system guided by a narrative of inclusion and recognition of the other-as-self, must create and refine its operational practices, procedures, and norms with the goal that individual providers solicit each patient's story. Doing so serves to avoid stereotyping and assumptive thinking that retard access to the other. Also, it should be recognized that a person can belong to, or identify with, more than one cultural group, depending on the situation or context. An individual, thus, can possess multiple identities that are revealed through her story.[34] Therefore, it is difficult to construct a universal approach that effectively categorizes or assesses people by group. It is misleading, presumptive, and paternalistic to believe and expect that people from the same group will all necessarily act alike.[35] Narrative Authority serves as a means to overcome the deficiencies of this limited thinking and helps to broaden the horizon of medical practices.

Understanding the importance of individual identity demands recognizing the identity and behavioral nuances a person embraces and acts upon at the *personal*, *group*, and *universal* levels. There is little doubt that people from similar backgrounds act, generally, in similar ways—they share similar narratives. However, it is essential to recognize this but not be blinded by it.

A patient may well act differently than another from the same group in very similar situations. Each person, by nature of her humanness, strives to construct, define, and articulate an individual self-identity that must be recognized and respected.[36] True cultural sensitivity implies focused attention on and commitment to the personal level, while, at the same time, understanding the role the group and universal levels play. However, access to the interconnected levels most often becomes apparent by focusing attention on the individual and her storied autonomy. Developing practices which direct the practitioner to concentrate first on the individual help to ensure that the system acts as a trustworthy other-oriented caregiver rather than a mere client-provider healthcare facility.[37]

12.2.3 Imposition and Effect of Western Colonialism

Globalization and multiculturalism present a fundamental challenge for the Western medical practices. Principlism is often criticized because of its Western world biases, which can result in cultural insensitivity. The Western viewpoint is founded

[34] Padela, Aasim I., Aisha Y. Malik, Farr Curlin, and Raymond De Vries. "[Re]considering Respect for Persons in a Globalizing World." *Developing World Bioethics* 15, no. 2 (2014): 98–100.

[35] Derald Wing Sue, "Multidimensional Facets of Cultural Competence," *The Counseling Psychologist* 29, no. 6 (2001): 790–793.

[36] Somnath Saha, Mary Catherine Beach, and Lisa A. Cooper, "Patient Centeredness, Cultural Competence and Healthcare Quality," *Journal of the National Medical Association* 100, no. 11 (2008): 6–8.

[37] Derald Wing Sue, "Multidimensional Facets of Cultural Competence," *The Counseling Psychologist* 29, no. 6 (2001): 792–798.

on the belief that the objective healthcare ethics principles or their interpretation cannot and must not vary from country to country. This belief is grounded on the foundational tenet that modern medicine is scientifically descriptive and unchangeable within a given period.[38] Based on this Western worldview—this guiding belief—scientists have developed bioethics theories, which are unquestioningly assumed to apply to and appropriate for any culture or situation.

For example, one of the four principles of bioethics is patient autonomy. The two underlining criteria of autonomy—*liberty and agency*—reveal that an autonomous person is not under the control of someone else: she can openly and freely present the choices she prefers.[39] This principle stirs great debate because the notion of radical autonomy is a fundamental component of the Western worldview, whereas, in the East, the majority of people understand themselves through a narrative that privileges communal decision-making and interdependence. Similarly, the understanding of and the near fanatical demand for informed consent takes center stage in Western medical ethics, while, ironically, developing countries fight for access to healthcare in any form, where the worry of informed consent, while important, seems trivial. This discrepancy exemplifies that the belief in universalism prioritizes Western healthcare ethics and overlooks or mistakes the possibility that there may be a more broadly defined, globally inclusive form of healthcare ethics. Thus, it is time to investigate the robust sway that Western-based assumptions exerts on 'biopolitics' and conscientiously shape a global, inclusive, and diverse 'biodemocracy.'[40]

For instance, a widespread practice in Pakistan is to defer to the physician's knowledge and authority to make a medical decision. In many cases, families and patients routinely relinquish medical decisions to the doctor's authority, his expertise. There is an unquestioned social trust integrated into the community narrative: physicians are considered as healing agents. Research indicates that a significant number of patients normatively abdicate decision-making authority to the physician. Concerning Western medical ethics, this would likely be interpreted as mere physician paternalism. However, in Pakistan, such actions are understood as respecting the individual's autonomy and are therefore ethically justified and valued within this cultural context, within this particular sense-making narrative. In this situation it is not paternalism but patient centrism.[41]

It is hard to apply Western individualism in Pakistan because people there believe, normatively and pre-reflectively, in interdependency. One need not be in Pakistan or outside the United States to experiences this. A similar situation occurred in Michigan concerning an Arab family and the decision to remove a loved one from

---

[38] Sirkku K. Hellsten, "Global Bioethics: Utopia or Reality?," *Developing World Bioethics* 8, no. 2 (2008): 71–73.

[39] Tom L. Beauchamp and James F. Childress, *Principles of biomedical ethics*, 7th ed. (New York: Oxford University Press, 2013), 101–105.

[40] Sirkku K. Hellsten, "Global Bioethics: Utopia or Reality?," *Developing World Bioethics* 8, no. 2 (2008): 81.

[41] Padela, Aasim I., Aisha Y. Malik, Farr Curlin, and Raymond De Vries. "[Re]considering Respect for Persons in a Globalizing World." *Developing World Bioethics* 15, no. 2 (2014): 98–104.

life support. The Arab family deferred the decision-making authority to the doctor. It took some time and a bit of persuasion for the Western doctor to understand that the family was not abdicating or avoiding choice-making. This was how the family chose to make the decision within their cultural norms. Generally, Arabs are as communal as Pakistanis. They also tend to believe in the positional authority of the physician. These examples highlight the fact that it is a mistake to presume and assume the global prevalence of the Western norm of autonomy in an ever diverse, intercultural world. In fact, in such increasingly diverse and pluralistic settings, it is culturally insensitive and merely unethical to privilege a Western perspective and assume its universality.[42]

The prevalent Western world bias in medicine creates confusion and, in fact, overlooks the possibility of a broader, more inclusive ethics. The Western notion of autonomy can in many cultures strip a patient of her personhood or self-identity. Therefore, Western world conceptions of autonomy and identity must be revised and modified. Respecting the normative understandings of personhood in different societies and cultures, and, in fact, among different individuals within similar contexts, should be championed as the universal standard in a cross-cultural healthcare system.[43]

Since Western beliefs and business models continue to skew bioethics models, it is must be recognized that it is no longer wise—or good business—to impose these conceptions onto the rest of the world. The understanding of ethics has grown. True ethics demands respecting each individual's identity, regardless of setting or cultural background.[44] Knowing their story helps us do so. The notion of bioethics must expand by adding nuance, and by showing that it is only possible to universalize common moralities by recognizing the specific situation and the social context.[45] This can be accomplished by embracing the local moralities and values of different cultures—by understanding the shaping forces of the local narrative. Doing such also requires a re-visioning of the unquestioned privilege given to the Western view in the normative and narrative practices shaping the healthcare system.

In fact, there are many non-Western medical practices around the world which are more normatively holistic than the process- and efficiency-oriented model of Western medicine. They arise from a different narrative understanding of the world. Examples can be found in Japan, Tibet, Korea, as well as in Ayurveda and Unani healing techniques. These approaches to medicine rely on an integrated notion of health and well-being that extends beyond responding to a symptom or wound. These practices of medicine fundamentally narratively embrace and respect the other dimensions of life, beyond the physical, which collectively encompass well-being. Conceptually

---

[42] Padela, Aasim I., Aisha Y. Malik, Farr Curlin, and Raymond De Vries. "[Re]considering Respect for Persons in a Globalizing World." *Developing World Bioethics* 15, no. 2 (2014): 105–106.

[43] Padela, Aasim I., Aisha Y. Malik, Farr Curlin, and Raymond De Vries. "[Re]considering Respect for Persons in a Globalizing World." *Developing World Bioethics* 15, no. 2 (2014): 104–106.

[44] Padela, Aasim I., Aisha Y. Malik, Farr Curlin, and Raymond De Vries. "[Re]considering Respect for Persons in a Globalizing World." *Developing World Bioethics* 15, no. 2 (2014): 99–100.

[45] Tom L. Beauchamp and James F. Childress, *Principles of biomedical ethics*, 7th ed. (New York: Oxford University Press, 2013), 10–25.

and narratively, these practices address and respond to well-being as a balance and intersection of the body, mind, and spirit. By disregarding or ignoring these Eastern narrative-based medical practices and perspectives, Western medicine reveals its insensitivity to other worldviews and other guiding narratives. Western medical biases will change only when organizations shift their perspective and revise their narratives to embrace and value the different medical practices and perspectives. Open-mindedness and multidisciplinary approaches will serve to integrate or assimilate the various medical practices and loosen the tight hold the Western perspective has on bioethics. The time is now to understand the foundational role narrative plays in shaping healthcare organizations. Understanding and welcoming the narratives that give rise to holistic and integrated care practices is pivotal if healthcare hopes to provide global-based care. Global-based care embraces healthcare as a human right. Thus, the standard care involved addresses, embraces, and responds to individual and cultural differences and integrates the preferences of each individual and her culture in the care delivery process.[46]

Developing a cultural sensitivity and global perspective to bioethics provides solid direction and firm grounding from which to address the many ethical dilemmas apparent in our ever-flattening, global society.[47] As a result, I suggest, Narrative Authority is the key, guiding principle to guide any organization to offer excellent and ethical health care in this new cross-cultural world.

12.3 Knowledge and Insights to Become Culturally Competent Healthcare

To become culturally competent, a healthcare system must design systems and procedures that reveal each patient's story. For this to occur an organization must revisit or re-narrate its organizational story, which serves to guide the necessary structural changes that, in turn, open a cultural space where each individual provider can be more attentive to and respectful of the patient's story. Likewise, each patient's story/profile reveals the person's race, gender, ethnicity, economic status, educational level, social status, fundamental beliefs, and likely her future aspirations and hopes. Knowing the person's profile—her story—provides insight into the individual's conceptualization of illness and her explanatory model of illness.[48] To become a culturally competent, patient-centric, and globally competitive healthcare provider, the system must develop a clear organizational narrative which articulates the importance of and belief in other-oriented, storied-driven bioethics. Narrative

[46] "Globalization of Healthcare," *Global Advances in Health and Medicine* 1, no. 2 (2012): 5–7.

[47] Thomas L. Friedman, *The World Is Flat: A Brief History of the Twenty-first Century* (Brantford, Ont.: W. Ross MacDonald School Resource Services Library, 2009), 308–309.

[48] Leigh Turner, "From the Local to the Global: Bioethics and the Concept of Culture," *Journal of Medicine and Philosophy* 30, no. 3 (2005): 305–309.

Authority offers the perspective and tools to articulate this narrative and to identify and implement subsequent and necessary structural and best-practice changes.

12.3.1 Understanding the Patient-Level Disparities

When a healthcare system provides services for a patient, it is imperative that the patient is seen as a *person* with a socially based, story-informed identity. This is an identity replete with experiences of and beliefs and fears about certain diseases. It is important to see the patient in an *as-individual* manner. Doing so acknowledges a patient's hopes, fears, and concerns, and offers *other*-centric healthcare that fosters a partnerships with the patient.[49] When practitioners work in conjunction with and through a patient's story they can understand her perspective and, in turn, enhance the quality of the healthcare provided. This narrative-driven, other-centric approach, by nature, serves to promote cultural sensitivity and build cultural competence systemically.[50]

However, the need to embrace and shape other-oriented healthcare systems is urgent. The diversity and disparities among patients grow at an increasing rates. More and more frequently, majorities encounter minorities. As a result individuals often face inequities and feel alienated, isolate, and alone. This, in turn, affects their assimilation or participation in the broader narrative. Often, they are left to feel as spectators, outsiders, rather than welcome participants. The sources of such inequalities are revealed via historically engrained norms and are generally supported and enabled by status-quo bureaucratic processes.[51] The healthcare system, coupled with the common, unquestioned, and status-quo narrative, drives and perpetuates such disparities.[52] For instance, ethnic minorities receive less treatment for diagnoses of depression and anxiety than do the white majorities because of the underlying and culturally ingrained belief that minorities—*they*—are less sophisticated and tend to misuse drugs.[53] Furthermore, due to socioeconomic disparities, there are significant access barriers to health care among marginalized groups.[54]

---

[49] N. Oflynn and S. Staniszewska, "Improving the experience of care for people using NHS services: summary of NICE guidance," *BMJ* 344, no. Mar16 1 (2012): 48–50.

[50] Somnath Saha, Mary Catherine Beach, and Lisa A. Cooper, "Patient Centeredness, Cultural Competence and Healthcare Quality," *Journal of the National Medical Association* 100, no. 11 (2008): 6–8.

[51] Laurie M. Anderson et al., "Culturally competent healthcare systems," *American Journal of Preventive Medicine* 24, no. 3 (2003): 68–70.

[52] Leslie J. Blackhall et al., "Ethnicity and attitudes towards life-sustaining technology," *Social Science & Medicine* 48, no. 12 (1999): 1781–1783.

[53] Laurie M. Anderson et al., "Culturally competent healthcare systems," *American Journal of Preventive Medicine* 24, no. 3 (2003): 68–70.

[54] Laurie M. Anderson et al., "Culturally competent healthcare systems," *American Journal of Preventive Medicine* 24, no. 3 (2003): 68–70.

Ethnic disparities are pervasive in healthcare even when adjusting for socioeconomic status. There are two consequences of the disparities. One, the performance of the hospitals, and, two, differential treatment patients receive within the hospital. Hospitals strive to garner high performance ratings. In this quest to measure, rate, and make the process ever more quantitatively efficient, a healthcare system often overlooks the qualitative components that give rise to excellent, other-oriented, ethical healthcare. Hospitals often neglect the narrative of care—the narrative of illness—and often its own guiding, healing organizational narrative to improve the patient's life. And in the second instance, the hospital supplies different treatment to different ethnicities because of stereotypes, unquestioned assumptions, and lack of cultural competence. This raises the chicken-or-the-egg question whether ethnic minorities receive inadequate care because of their ethnicity or because of the setting in which they are receiving care.[55]

These patient-level disparities occur in many segments of the system. For instance, a common form of disparity occurs when a hospital discharges a patient. Hospitals that have more resources tend to spend more time to counsel and prepare patients for discharge. However, hospitals with limited resources usually have less staff designated to counsel patients who are going to be discharged. In many cases, minorities are served by a lesser-quality hospital with fewer resources. Beyond limitations of financial and staff resources, language also plays a critical role in creating patient-level disparities. Often staff and nurses feel less comfortable or confident interacting with a non-English-speaking patient. Because of this, miscommunications and misunderstandings can occur and patients leave the hospital with incomplete knowledge and mistaken assumptions about care. Cultural and linguistic training coupled with sufficient organizational resources must be provided in the quest to ensure cultural competence and ethical care. Patients who are properly discharged from the hospital is one indicator of an organization that builds trustworthy partnerships.[56]

Besides discharging patients, disparities are also widespread in preventive care. For instance, studies reveal that African American and Hispanic patients are less likely to receive influenza or pneumococcal vaccines than are White patients. Surprisingly, more than one-third of all minorities remain unvaccinated in comparison with White patients. Lack of access to primary care and an incomplete understanding of vaccinations are causes for such disparities in preventive care.[57]

---

[55] Romana Hasnain-Wynia et al., "Racial and Ethnic Disparities within and between Hospitals for Inpatient Quality of Care: An Examination of Patient-Level Hospital Quality Alliance Measures," *Journal of Health Care for the Poor and Underserved* 21, no. 2 (2010): 629–633.

[56] Romana Hasnain-Wynia et al., "Racial and Ethnic Disparities within and between Hospitals for Inpatient Quality of Care: An Examination of Patient-Level Hospital Quality Alliance Measures," *Journal of Health Care for the Poor and Underserved* 21, no. 2 (2010): 630–634.

[57] Romana Hasnain-Wynia et al., "Racial and Ethnic Disparities within and between Hospitals for Inpatient Quality of Care: An Examination of Patient-Level Hospital Quality Alliance Measures," *Journal of Health Care for the Poor and Underserved* 21, no. 2 (2010): 643–644.

All these disparities can be overcome through a commitment to the transformative power of Narrative Authority. Narrative Authority offers the power to uncover the social-economic background of the patient and provide an approach by which to comprehend, respect, and value a patient's worldview. This worldview guides the provider to walk into the patient's shoes. This occurs by listening to the other's story, through which the practitioner sees, understands, and *is-with-the-other-as-a-person,* not as a patient or customer—but as *caregiver,* not as an institutional agent.[58]

The road to change is different for every healthcare system. Each has its own story. Articulating this story opens the space for the culture changes and process improvements necessary to supply progressive and revolutionary healthcare. Regardless the particular situation, Narrative Authority serves as the vehicle which helps to make this journey easier. Narrative Authority works as a magnet to bridge the gap between what is expected and what is actual. Narrative Authority unifies diverse fields of thoughts to help understand the varied cultural practices within the organization.[59]

However, we live in a time when the healthcare system is designed within a performance-based structure, and thus serves two opposing masters: profitability and patient care. Over the past 20 years, healthcare systems in the USA have changed dramatically in the direction of profitability, which generates bureaucracy and system rigidity that foments mistrust and uncertainty for patients. Healthcare providers have established roles as profit-making systems rather than human healing agents. The system has become a cost-conscious bureaucratic marketplace which is indifferent to the patient's suffering.[60] In such settings, practitioners often have little time, and often less impetus, to see the patient as a person—the setting is designed to see the patient only as a *customer*. A physician, in the face of his rush to meet organizational dictates, is not allotted enough time to take a patient's oral history or investigate her story. Efficiency drowns out the illuminative power of narratives.[61] Not knowing the patient's history retards optimal, ethical, other-oriented care. Optimal care rests on understanding the patient's storied history, which provides the physician with a sense of the narrative coherence that shapes the patient's life.[62] Therefore, understanding a patient's perspective—her story—helps to avoid patient-level disparities. Story gives the provider access to her culture and normative surroundings, which commonly determine a patient's sense of wellbeing. Therefore, if the healthcare system wants to provide excellent and ethical care, it

[58] Clive Baldwin, "Narrative, Ethics and People with Severe Mental Illness," *Australian & New Zealand Journal of Psychiatry* 39, no. 11–12 (2005): 1025–1027.

[59] Rita Charon, *Narrative Medicine: Honoring the Stories of Illness* (Oxford: Oxford Univ. Press, 2008), 9–12.

[60] Rita Charon, *Narrative Medicine: Honoring the Stories of Illness* (Oxford: Oxford Univ. Press, 2008), 3–4.

[61] Donna L. Washington et al., "Transforming Clinical Practice to Eliminate Racial-Ethnic Disparities in Healthcare," *Journal of General Internal Medicine* 23, no. 5 (2008): 688–89.

[62] Clive Baldwin, "Narrative, Ethics and People with Severe Mental Illness," *Australian & New Zealand Journal of Psychiatry* 39, no. 11–12 (2005): 1023–1026.

must design and structure its systems and practices in ways that provide physicians the appropriate time and flexibility to get to know the patient's background—to hear her story.[63]

As in every other aspect of society, there are gender disparities in healthcare. Gender discrimination and gender inequality in healthcare are not always immediately evident. Gender-related obstacles to creating patient-centric healthcare are often blurred by or hidden within other aspects of the organization's culture and normative practices.[64] Females are generally just not considered as important as males because they typically produce less income and are seen as less central to the everyday workings of society. This predisposed and stereotypical position proves to be disadvantageous for females. They have less bargaining power in the roles they can assume in the workplace, family, or community.[65] In countries like India, for instance, where poverty, hierarchy, and patriarchy are the prevailing norms, women generally are confined to household duties and have less connection with or access to the broader society. Even in Canada, a developed country, research reveals that women receive less treatment than men for chronic diseases.[66]

Therefore, gender plays direct, indirect, and often unrecognized roles in the patient-physician relationship in healthcare. These inequalities and disparities can be addressed and overcome by creating an environment and practices that serve all actors involved in healthcare as a human being—an other-as-self with equal rights and reciprocal obligations. The adverse effects of gender discrimination on the patient-physician relationship can be overcome through a commitment to the transformative power of Narrative Authority. This transformation occurs by embracing the other's story. Through her story one sees, understands, and is-with the other as a person not as patient, as caregiver not as an institutional agent.[67] Therefore, if the system recognizes the patient-level disparities within society, it can respond by developing practices that value and capture patients' stories and worldviews.

As has been stated above, the field of healthcare is in flux and is confronted with enormous opportunities to embrace change, expand, and enhance its quality and reach. One consistent inhibitor is that bias, stereotyping, and unquestioned assumptions are often not well addressed, and change is arrested. With the advancement of technology and medical science, the story and narrative which ground our beliefs, must change to promote and demand—as both ethical and business imperatives—inclusion, diversity,

---

[63] Somnath Saha, Mary Catherine Beach, and Lisa A. Cooper, "Patient Centeredness, Cultural Competence and Healthcare Quality," *Journal of the National Medical Association* 100, no. 11 (2008): 5–8.

[64] Robert M. Veatch, *Patient, heal thyself: how the new medicine puts the patient in charge* (Oxford: Oxford University Press, 2009): 10–13.

[65] Leslie J. Blackhall et al., "Ethnicity and attitudes towards life-sustaining technology," *Social Science & Medicine* 48, no. 12 (1999): 1785–1787.

[66] Simone Dahrouge et al., "An evaluation of gender equity in different models of primary care practices in Ontario," *BMC Public Health* 10, no. 1 (2010): 6–10.

[67] Robert M. Veatch, *Patient, heal thyself: how the new medicine puts the patient in charge* (Oxford: Oxford University Press, 2009): 10–13.

and commonality. Healthcare practice can and will only change when guided by a narrative that offers no place for disparities. The means to undo common bias and stereotyping is to shift the perspective on how patients and physicians understand, see, and embrace diversity. The goal is to become more individually and organizationally self-reflective. Imagining and articulating a guiding narrative that discloses a space where vulnerable populations and diverse individuals of all types are seen not as other, but as people; not as things, but humans.[68]

12.3.2 Explanatory Model of Illness and Conceptualization of Illness

Contemporary healthcare has spread beyond what was previously considered traditional medical practices. Due to globalization and multiculturalism, health practitioners must now be skilled, trained, and attuned to a patient's perspective, her perception of illness, and the narrative roots shaping her understanding of illness. Care is no longer only about diagnosing and treating disease, but also about finding the related contextual features that can affect an individual's or a family's wellbeing. Health care providers must understand the patient's beliefs about health to understand the patient and to respond in an other-oriented, culturally sensitive, ethical manner. Also, providers should note, acknowledge, and respect the interplay between the social and cultural influences of an individual to accurately understand the story-driven context of the illness. Care in a multicultural society expands the status-quo to embrace an orientation that respects the patient, communicates well and checks for understanding, is rich with active listening, acknowledges and embraces cultural differences, and regards the patient as a human other in need of care.[69]

Comfort and mental wellbeing are known to positively affect and transform physical wellbeing. Especially in healthcare, it is a must to have a therapeutic model of care to ensure that the patient actively participates in and contributes to the healing process. A therapeutic model of care that is driven by open-ended questions, reflective responses, and uses silences and pauses serves build a dialogue-based partnership between the care provider and patient. Since we live in a multicultural society, not every therapy or procedure will work for all. Care providers should keep this in mind when practicing a therapeutic model of care. For instance, a therapeutic model of care might face challenges in a particular culture. E.g., Indian patients generally believe that a care provider has knowledge power and expect the provider

[68] Rita Charon, "Narrative Medicine," *JAMA* 286, no. 15 (2001): 1899–1900.

[69] Iain Graham and Linda Norman, "A Reflective Discussion: Questions about Globalization and Multiculturalism in Nursing as Revealed during a Student/staff Exchange Programme," *International Journal of Nursing Practice* 14, no. 3 (2008): 189–194.

to be prescriptive. In this instance the care provider should use a narrative approach to resolve the power distance and make the patient speak.[70]

Sustainable healthcare should always be conscious to satisfy both mental and physical wellbeing by ensuring a therapeutic model of care. However, that may only be possible when the care providers know how the patient perceives the illness: they have to know her story. The perception of illness and the explanatory model of illness are the way to understand how the patient and the patient's family see, understand, and are influenced by the illness.[71] For instance, consider a patient in the emergency room. The patient and the patient's family commonly experience higher distress levels than the average patient admitted in that hospital and require soothing and comforting care. It is not possible to provide the proper comfort care without understanding the patient's view of illness. Pain, suffering, and distress are conditions which are very difficult to measure, only through engaging with the narrative of the patient it is possible to understand the patient's or family's view of the illness.[72]

Each patient's explanatory model of illness and her conceptualization of illness is unique and differs even from those from the "same" background. A patient's story—which includes immigration history, a cultural perspective, a social network, a socio-economical history, and educational level—shapes the perspective of illness. Therefore, a healthcare system must develop creative and adaptable narrative-based protocols to determine the patient's sense of illness.[73] A patient's socioeconomic background and education also determine her conceptualization of illness. Even healthcare literacy defines a person's conceptualization of illness.[74] All these are pieces of her story—they tell the provider who she is and inform what she believes.

In this context, it is vital to state again that those directing global healthcare growth must recognize that Western worldviews are not universal. The population of the Western world is only a portion of the global population. Almost one-third of the world's population lives in emerging countries. People in these countries often rely on traditional healing or self-care. The views and the perceptions of those patients from cultures accustomed to traditional medicine will not change their viewpoints easily or quickly when they migrate to a Western country. For instance, there are groups who believe in guardian angels and hold the belief that illness is sent from God and must be healed by God. These people believe in the power of the supernatural. Their belief is not easily modified by the rationality of traditional

---

[70] M. C.-T. Tai, "Developing a Culturally Relevant Bioethics for Asian People," *Journal of Medical Ethics* 27, no. 1 (2001): 51–52.

[71] David B. Morris, "Narrative, Ethics, and Pain: Thinking with Stories," *Narrative* 9, no. 1 (2001): 60–63.

[72] Rita Charon, *Narrative Medicine: Honoring the Stories of Illness* (Oxford: Oxford Univ. Press, 2008), 4–6.

[73] J. Emilio Carrillo, "Cross-Cultural Primary Care: A Patient-Based Approach," *Annals of Internal Medicine* 130, no. 10 (1999): 830–832.

[74] J. Emilio Carrillo, "Cross-Cultural Primary Care: A Patient-Based Approach," *Annals of Internal Medicine* 130, no. 10 (1999): 829–831.

Western medical practices. Such patients will look for therapeutic communication that satisfies and aligns with their indigenous understanding of healing. Unfortunately, healthcare institutions often ignore or overlook the centrality of indigenous beliefs that strongly shape the explanatory model of the patient's illness. Therefore, healthcare organizations must value indigenous medical practices and the narratives that give rise to those beliefs to properly understand the illness of the patient in a multicultural setting.[75] A means to see or gain access to a person is through her story. Narrative Authority guides caregivers to solicit those stories as a pathway to provide culturally competent care. Cultural competency therefore helps the provider find her way to the patient's values, expectations, fears, and beliefs as a part of the treatment process.

However, the economic pressure to be efficient can often lead to time-limited visits, which further push practitioners to compartmentalize, categorize, and stereotype "patients" of the same culture or same group. The practitioner often assumes that every person of a particular group falls into "that category" and thus acts and believes similarly.[76] Generalization assumes a common ground for each individual culture. By understanding cultural variance, the system's processes and, as a consequence, the individual provider can embrace the importance of implementing specific and well-designed best practices for each individual scenario.[77] Therefore, the system cannot ethically generalize or prescribe a single way to address a group of people. Doing so creates stereotypes and biases. In such situations, the practitioner stops listening and misses the nuance and the insights the patient offers through her story. These are instances where the organization abandons its commitment to care.

The organization should design and develop individual practices and procedures that eliminate biases and blind spots. Regardless of the situation, a physician must be placed in a position to consistently recognize and account for the story-based foundation of the patient's explanatory model of illness.[78] To gain a proper diagnosis, it is highly recommended that the practitioner collects details about the patient's explanatory model of illness, and remains aware of the fact that each individual sees the world through her own storied eyes: we see through the culture and narrative we inherit.[79] Each patient's view of care and pain is particular and nuanced, which is

---

[75] Mesfin H. Kahissay, Teferi G. Fenta, and Heather Boon, ",," *BMC Public Health* 17, no. 1 (2017): 1–5.

[76] J. Emilio Carrillo, "Cross-Cultural Primary Care: A Patient-Based Approach," *Annals of Internal Medicine* 130, no. 10 (1999): 829–831.

[77] Somnath Saha, Mary Catherine Beach, and Lisa A. Cooper, "Patient Centeredness, Cultural Competence and Healthcare Quality," *Journal of the National Medical Association* 100, no. 11 (2008): 7–9.

[78] J. Emilio Carrillo, "Cross-Cultural Primary Care: A Patient-Based Approach," *Annals of Internal Medicine* 130, no. 10 (1999): 830.

[79] Karori Mbugua, "Respect for Cultural Diversity and the Empirical Turn in Bioethics: A Plea for Caution," *Journal of Medical Ethics and History of Medicine* 5 (2012): 2–5.

why it is imperative that a physician understands her story and communicates effectively with the patient to avoid stereotyping or assumptive thinking.[80]

Veatch makes clear the difficulty of creating a universal framework for medical decisions, given different religious worldviews and philosophical beliefs. It is simply inappropriate to define *one* correct set of beliefs and values as *the* guide to the practice of medicine.[81] Knowing a patient's explanatory model of illness, exploring a patient's agenda, and contextualizing illness help physicians act in a more appreciative, assistive manner.[82]

Each person's manner of conceptualizing illness is specific because it is influenced and grounded by her explanatory and story-based model of illness.[83] It is routine to diagnose the biological understanding of an illness, but it is challenging and more nuanced to access the psychological and story-based components that shape a patient's conceptualization of the illness. The ideal way to see the patient's illness or beliefs is to have the "physician enter the patient's world, to see the illness through the patient's eye—to know her story."[84] Because, each person holds different normative beliefs, which affect their conception of the illness, the physician's autonomy and his perception of the patient can result in a different outcome for the same illness.[85]

A physician's knowledge and position power often inhibits the practitioner from making an accurate diagnosis of an illness. A physician possesses technical insights and knowledge that the patient does not have. This gives her power over the patient. Therefore, the physician must be attuned to the cultural perspectives offered by the patient. A physician should never initiate treatment from an assumed understanding or "knowing position" because the patient's conceptualization of illness determines and shapes the type of care needed.[86] Understanding and being open to the patient's conception of illness is important if the physician is to provide a proper diagnosis. Therefore, making assumptions about educational levels or other categorizations erects obstacles to achieving an effective and ethical other-oriented healthcare system. By being empathetic and attuned to the other's story, it is possible for a

[80] Somnath Saha, Mary Catherine Beach, and Lisa A. Cooper, "Patient Centeredness, Cultural Competence and Healthcare Quality," *Journal of the National Medical Association* 100, no. 11 (2008): 3–8.

[81] Robert M. Veatch, *Patient, heal thyself: how the new medicine puts the patient in charge* (Oxford: Oxford University Press, 2009), 14–17.

[82] J. Emilio Carrillo, "Cross-Cultural Primary Care: A Patient-Based Approach," *Annals of Internal Medicine* 130, no. 10 (1999): 829–834.

[83] Mary Curry Narayan, "Six Steps toward Cultural Competence: A Clinician's Guide," *Home Health Care Management & Practice* 14, no. 1 (2001): 42–44.

[84] Somnath Saha, Mary Catherine Beach, and Lisa A. Cooper, "Patient Centeredness, Cultural Competence and Healthcare Quality," *Journal of the National Medical Association* 100, no. 11 (2008): 1–5.

[85] Mary Curry Narayan, "Six Steps toward Cultural Competence: A Clinician's Guide," *Home Health Care Management & Practice* 14, no. 1 (2001): 43–44.

[86] E. A. Berlin and W. C. Fowkes, Jr., "A Teaching Framework for Cross-Cultural Health Care. Application in Family Practice," *West J Med* 139, no. 6 (1983): 934–938.

physician to help a patient imagine the "next chapter" of her life and thus envision a fulfilling and meaningful future, which the patient has yet to see or believe.[87] Again, knowing or helping to articulate the patient's story is key.

There is no evidence that shows that the physician alone can best judge or guide the well-being of the patient. Because a patient's well-being is not only defined by medical or physical elements. Well-being is also greatly influenced by the combination of social, psychological, and religious beliefs and norms that the patient holds and practices.[88] A physician who enters a professional relationship with the client, believing that he or she already has the answer or knows best, does a disservice to the individual and acts in ways that overshadow the patient's autonomy, which leads to forced or uninformed medical practices. In short, in such cases, the physician serves the patient's benefit from a limited and self-serving viewpoint.[89] The organization's narrative, if genuine and grounded in healing, provides the sense-making and the urgency to design practices that provide the practitioner the flexibility to spend the appropriate amount of time to understand the patient's story and worldview.

Therefore, to gain a proper diagnosis and to determine a patient's acceptance level, it is highly recommended that the physician communicate in ways the patient understands, and then offers culturally clear options and choices that lead to consensus. This process is often difficult when the patient is incompetent. For an incompetent patient, the physician should communicate with the family and understand the patient's history to meet the needs of the patient.[90]

12.4 Actions to Create a Culturally Competent Healthcare System by Practicing Narrative Authority

The healthcare system needs to be built on the belief in cultural competency. The organization must construct its best practices to align with and be attuned to the notion of cultural competency, and the embrace of the other as a human being. Therefore, to realign the organization's perspective and to guide it towards becoming culturally competent, senior leadership should adopt Narrative Authority as the organization's fundamental, guiding principle. The system should expect every stakeholder to approach, communicate, and collaborate with every other stakeholder in an other-oriented, story-based, and dialogue-grounded manner. The story should be consistently communicated throughout all levels of the organization with

[87] Robert M. Veatch, *Patient, heal thyself: how the new medicine puts the patient in charge* (Oxford: Oxford University Press, 2009): 61–63.

[88] Robert M. Veatch, *Patient, heal thyself: how the new medicine puts the patient in charge* (Oxford: Oxford University Press, 2009): 30–39.

[89] Tom L. Beauchamp and James F. Childress, *Principles of biomedical ethics*, 7th ed. (New York: Oxford University Press, 2013): 220–225

[90] Meg Zomorodi and Barbara Jo Foley, "The nature of advocacy vs. paternalism in nursing: clarifying the 'thin line,'" *Journal of Advanced Nursing* 65, no. 8 (2009): 1748–1751.

the aim to be an ethical and culturally sensitive care provider. Narrative Authority provides the perspective from which to interact and understand each individual's story, improve and strengthen best practices, and, as a byproduct, make the system highly responsive, ethically justified, and globally competitive.[91]

12.4.1 Narrative Authority

A narrative provides a context where ethics make sense. Ethics arise through the moral imagination revealed through the narrative act. Without knowing someone's narrative it is impossible to judge the morality of a certain behavior or act. Ethical decision-making is not straightforwardly choosing good over evil. In fact, ethical decision-making is honoring one value at the expense of another value. A narrative determines which value is most suitable for a particular situation because a narrative is inseparable from individual acts which are determined by values.[92] The clarity and intent of a narrative helps form substantial ethics. A narrative gives sense-making, meaning, and urgency to a shared context which then defines and shapes moral behavior. The narrative heart of ethics calls the story, the teller, the listener, and the witness into the context— the space it opens—and immerses all involved into a co-created process of being-with.[93] The narrative aspect presents phenomenology from the deontological perspective. Deontology claims that acts of moral decision-making matter whereas a phenomenology of narrative reveals the underlying ground for intersubjectivity itself—discovering that the narrative act requires *the other* and is hence the experience of shared being. The narrative act reveals ethics to be the relationship of human connection. From here, processes and ways of moral and ethical decision-making arise.[94]

A narrative is a guiding story that configures understanding and context. As such, it serves to bring two people together, to be-with. In this space the two (or more) find access to collaborative, shared, and ultimately creative discourse, and a means to articulate and implement possible outcomes. This occurs because stories help a person understand another person's perspective from various angles. The story unfolds the depth of someone's viewpoint through context. The story helps the hearer, *see*. The story also helps the listener understand the reasons, grounding, and meaning behind the other's identity, and her presence in her world. And the story

[91] Clive Baldwin, "Narrative, Ethics and People with Severe Mental Illness," *Australian & New Zealand Journal of Psychiatry* 39, no. 11–12 (2005): 1025–1028.

[92] David B. Morris, "Narrative, Ethics, and Pain: Thinking with Stories," *Narrative* 9, no. 1 (2001): 68–71.

[93] Clive Baldwin, "Narrative, Ethics and People with Severe Mental Illness," *Australian & New Zealand Journal of Psychiatry* 39, no. 11–12 (2005): 1022–1024.

[94] Adam Zachary Newton, *Narrative Ethics* (Cambridge, Mass.: Harvard Univ. Press, 1997), 5–12.

always serves to bring two people together in the intersection and recognition of their shared humanness.[95]

In this chapter 'story' and 'narrative' are used interchangeably. However, there is a subtle difference between both concepts. Narrative indicates the structural, radically social way in which human beings understand themselves and others,[96] because we are always in relation with others in a shared context. Story is a particular expression of a narrative that illuminates the sense making and commonality of shared, intelligible, and meaningful context.[97] Since story reveals the narrative there is an obvious overlap between both concepts; this why the two terms are used interchangeably here.

Stories reveal how the particulars of health are related to the wholeness of wellness. An understanding of an individual's wellness can be best assessed by knowing and understanding her narrative, her story. The story allows a person to reveal herself—to show herself—to the other.[98] The story discloses what is significant to her, what her dream are, what challenges she faces, how she sees herself in the throes of her current predicament, and what her beliefs and guiding principles are. Her story reveals her worldview. Her story presents her as a person in the presence of another person who, too, has a story. Her story provides an intersection through which the physician, who is listening, sees her not as patient *there*, but as another-as-him *here*. The story helps the physician understand a patient-as-person, and enables her, thus, to act as healer, not provider.[99]

Yet, the guiding narrative is the light that remains in many ways hindered. The term, narrative, carries three different notions: (1) the story only, (2) the text of the story alone, or (3) the narrative act. I use the third connotation in this paper when describing Narrative Authority. *Narrative,* here, means the act of initiating and revealing both the story and the process of the narrative act. On the other hand, *Authority* means the power of and the guideline for leading a process. Here, the word authority does not represent autonomous power but the collective and shared power that emerges through the narrative act.[100]

Therefore, together, the two words, *Narrative Authority*, describe the foundational structure of human being. It denotes the interconnected and radically shared storied-space in which humans find sense making between the teller, listener, and witness. Here, in the midst of the space, story reveals the ethics—the context where

[95] Martha Montello, "Narrative Ethics," *Hastings Center Report* 44, no. S1 (2014): 3.

[96] A. H. Jones, "Narrative Based Medicine: Narrative in Medical Ethics," *BMJ* 318, no. 7178 (1999): 253–254.

[97] Colin Robertson and Gareth Clegg, *Storytelling in Medicine: How Narrative Can Improve Practice* (Boca Raton: CRC Press, Taylor & Francis Group, 2017), 3–5.

[98] Clive Baldwin, "Narrative, Ethics and People with Severe Mental Illness," *Australian & New Zealand Journal of Psychiatry* 39, no. 11–12 (2005): 1022–1024.

[99] Rita Charon, *Narrative Medicine: Honoring the Stories of Illness* (Oxford: Oxford Univ. Press, 2008), 97–99.

[100] Adam Zachary Newton, *Narrative Ethics* (Cambridge, Mass.: Harvard Univ. Press, 1997), 20–21.

responsibility to and obligation for the other are intertwined and inseparable. The responsibility-obligation act is twofold. On the one hand, it is an ethical duty to engage and reveal one's own story. On the other hand, it calls for the attention and openness to incorporate, welcome, embrace, and embody the other's story as a response. This process of simultaneously giving-and-receiving are lived acts of love. Narrative Authority is the process of giving oneself over to the foundational nature of the narrative act and to stand in the shared space where one sees the other as self. From here ethics arise—as the obligation to and responsibility for the other as self.[101]

Narrative Authority reshapes the normative assumptions concerning human relationships and makes a relationship inseparable from the shared space the story reveals, and the social and referential context in which it occurs. It is within this space that meaning, coherence, and intelligibility of the everyday occurs. Narrative Authority rests on the belief that all human relationships are radically social. Narrative Authority serves to revise the notion of the individual as separate and autonomous. Narrative Authority redefines individuality as socially shared and hence ethical by nature. The embrace of the narrative act is at once both the revelatory command of the obligation to self and the responsibility for the other. Narrative Authority helps individuals sense and understand that they are always and primordially socially connected to one another. There is no escape: this is the fabric of human being. Narrative Authority helps all participants re-cognize that being-in-conversation is to co-create a dialogue. Narrative Authority allows participants to embrace the normal motion of the everyday as a collection of ethical practices and beliefs which implicitly and explicitly honor the context of shared being.[102]

Narrative Authority helps to reveal the ethics of engaged presence. By exercising Narrative Authority, it is possible to understand the context of the concerning issues that the organization faces: a shared obligation to and responsibility for the patient and the society at large. Narrative Authority helps individuals and society form an engaged presence through the ubiquitous context of shared story, i.e. shared being. Of course, sometimes the patient's narrative can misguide the listener. This occurs when the listener is deaf to or dismissive of the patient's story and misses the nuances. The patient's narrative, in this case, can then misguide the "listener" to a wrong outcome or to a presumed conclusion. Often, providers over-hear or under-hear the patient's story. For instance, a patient may want to share something but the provider hears something else by missing the context or the nuances that her story provides. Also, the provider sometimes manipulates the narrative to fit a preconceived notion he or she holds. Understanding the narrative of others without manipulating it is the art of active and engaged listening. Therefore, Narrative Authority itself calls for a broader spectrum of individuality that is by nature social and ethical.[103]

---

[101] Adam Zachary Newton, *Narrative Ethics* (Cambridge, Mass.: Harvard Univ. Press, 1997), 14–20.

[102] Adam Zachary Newton, *Narrative Ethics* (Cambridge, Mass.: Harvard Univ. Press, 1997), 20–25.

[103] *Stories and Their Limits: Narrative Approaches to Bioethics*(Florence: Taylor and Francis, 2014), 1–16

12.4.2 Prioritize the Patient

In cross-cultural healthcare, Narrative Authority is a means to solve ethical dilemmas raised by globalism and multiculturalism. Relying on the power of Narrative Authority, an organization can ensure that processes, procedures, and practices guide stakeholders to see a patient as a person with an individual identity. Every ethical encounter should be approached person-to-person, where human rights prevail.[104] To be culturally competent and to act responsively, an organization should rely on practices that exercise Narrative Authority as normative. Narrative practices provide an accurate picture of what is going on in the patient's lifeworld. Because humans are fundamentally social, we are connected through and are constituted by and through narratives. By following Narrative Authority, caregivers can become more fully and meaningfully engaged with patients and more meaningfully engaged in their stories.[105]

Illness is isolation. Often patients feel isolated when they find themselves confronting the reality of disease. In systematic healthcare, there are many instances when the nuance and clarity of a patient's story are incompatible with a standardized, computerized, checklist, work-flow process. In this instance, human interaction is turned to abstraction—a faux reality—and leads to isolation and poor "doctoring." This is particularly common in cases of dementia, patients confined to their homes, and elderly patients.[106]

Structured "appointment windows" and length-of-service measurements hinder the practitioner's ability to delve into the worldview of a patient. The practitioner is forced into a time-efficient, mechanistic process that undermines his obligations to the patient-as-human and causes the practitioner to forgo normative ethics. The patient-physician relationship becomes a *patient-institution* relationship rather than a *person-person* relationship. Therefore, an organization has to change or revise its narrative and its understanding of its fundamental model. It must transform from a machine model to a humanistic, communal model. Narrative Authority serves as a mediator to stimulate a dialogue between patient and physician with the intent to form a partnership.[107] Narrative Authority provides the foundation from which to revise and create practices and protocols that ensure the patient and physician build and maintain a partnership base on shared and intersecting stories.

---

[104] Somnath Saha, Mary Catherine Beach, and Lisa A. Cooper, "Patient Centeredness, Cultural Competence and Healthcare Quality," *Journal of the National Medical Association* 100, no. 11 (2008): 6–8.

[105] Clive Baldwin, "Narrative, Ethics and People with Severe Mental Illness," *Australian & New Zealand Journal of Psychiatry* 39, no. 11–12 (2005): 1022–1024.

[106] Regina Harrell, "Why A Patient's Story Matters More Than A Computer Checklist," NPR, November 17, 2013, accessed June 23, 2018, https://www.npr.org/sections/health-shots/2013/11/17/242366259/why-a-patient-s-story-matters-more-than-a-computer-checklist

[107] S. A. Buetow, "To Care Is to Coprovide," *The Annals of Family Medicine* 3, no. 6 (2005): 553–556.

Therefore, the structure of a healthcare organization should allow more "face time" with the patient to decrease the patient's stress and improve the patient's experience. A longer appointment window can help a physician capture the history of the patient and thus avoid or overcome biases and stereotyping. Time pressures force quick judgments, increase cognitive load, and result in more decisions founded on unquestioned assumptions. Therefore, rather than working for mere efficiency, the system should strive for effective care.[108] The healthcare system should not operate like a machine, instead it should act as an integrated social body with a narrative conscience.

The organization should develop and maintain a database to house the patient's information *and* stories. The system can, in turn, assess the data and create a reference for the physician. If a physician encounters a person from a culture with which he is not familiar, having the cultural information available in the database can help the care provider stay aware of his or her perspectives and unquestioned assumptions while administering care. For instance, the practice of a Navajo patient and her expectations could be in the database as knowledge for care providers who often do not attend to Navajo patients. This would help to avoid stereotype and biases and reduce the occurrence of disparities.[109]

Despite the numerous benefits of technology in the medical field, organizational structures and procedures dedicated to efficiency and profit distort the physician's gaze and he or she, hence, comes to see a customer, not a patient. For example, during a consultation, a physician keeps jotting case notes into a networked computer. Indirectly, both the patient and the physician communicate to the computer, to the calculative process of efficiency. The relationship is mediated. In the process of collecting "information," the physician is procedurally detached from making an emotional connection with the individual identity of the patient. Thus, he or she, in turn, loses the scope of normative ethics, narrative practices, and human connection.[110] Exercising Narrative Authority is fundamentally a moral response to the existential appeal of the patient, where the physician not only hears the patient's story but also tells her story to others involved in the decision-making at hand.[111] To become a culturally competent and responsive healthcare system, the voices of both the physician and patient must be heard, valued, and respected. By exercising Narrative Authority both the patient and the physician can experience the actual meaning of autonomy, which can be dependent, independent, and interdependent depending on the perspective and mood of a particular patient or setting. Narrative Authority helps identify the fundamental values of a person and

---

[108] Leslie J. Blackhall et al., "Ethnicity and attitudes towards life-sustaining technology," *Social Science & Medicine* 48, no. 12 (1999): 1781–1786.

[109] Mary Curry Narayan, "Six Steps toward Cultural Competence: A Clinician's Guide," *Home Health Care Management & Practice* 14, no. 1 (2001): 42–44.

[110] Mary Curry Narayan, "Six Steps toward Cultural Competence: A Clinician's Guide," *Home Health Care Management & Practice* 14, no. 1 (2001): 40–44.

[111] Clive Baldwin, "Narrative, Ethics and People with Severe Mental Illness," *Australian & New Zealand Journal of Psychiatry* 39, no. 11–12 (2005): 1024–1029.

thus helps to protect the person's identity.[112] Therefore, considering the ultimate goal of care, both the physician and the patient should act as partners—where Narrative Authority prevails.

Narrative Authority should, therefore, also guide the informed consent process. Patients are not properly escorted through the consent, because time constraints or predetermined structures make healthcare providers rush through the process. In many cases, patients give consent without understanding the treatment process. Time restricted processes often thwart autonomy by not providing proper attention to the patient to gain proper informed consent. This can be overcome by fostering culturally sensitive and culturally responsive narrative practices that promote ethical engagement and build client-provider partnerships. More so, through a culturally sensitive, dialogue-based, and story-bound partnership, the system allows the individual caregiver to understand and embrace autonomy as an obligation, and, subsequently, view consent as consensus. Thus, the care provider can understand the importance of properly guiding a patient through the consent process without rushing. Narrative Authority grounds this partnership. Narrative Authority guides the patient and physician to see the *other-as-same* and to find common connections through a story-based, dialogue-oriented, mutually shared relationship.[113] A dialogical and story-based interaction solicits the sincere viewpoints of and connects each stakeholder engaged in the conversation.[114]

12.4.3 Practice Effective Communication

Communication enhances patient satisfaction, provides coherence, and promotes healthy outcomes.[115] Effective communication connects understanding, meaning, and context between the patient and physician, and thus encourages cultural responsiveness and preserves cultural competency. Even though language and communication may appear to be "the same," they can be interpreted in remarkably different ways in what may seem similar contexts. Two people may speak the same language, but this fact alone does not guarantee effective communication—where both parties understand one another's positions and points of view. Even in a case where there is an English-speaking patient and an English-speaking physician, there may

[112] Godelief M.j. Mars et al., "Conceptualizing autonomy in the context of chronic physical illness: relating philosophical theories to social scientific perspectives," *Health:: An Interdisciplinary Journal for the Social Study of Health, Illness and Medicine* 12, no. 3 (2008): 336–340.

[113] Clive Baldwin, "Narrative, Ethics and People with Severe Mental Illness," *Australian & New Zealand Journal of Psychiatry* 39, no. 11–12 (2005): 1025–1027.

[114] Tineke A. Abma et al., "Inter-Ethics: Towards An Interactive And Interdependent Bioethics," *Bioethics* 24, no. 5 (2010): 253–255.

[115] J. R. Betancourt et al., "Defining Cultural Competence: A Practical Framework for Addressing Racial/Ethnic Disparities in Health and Health Care," *Public Health Rep* 118, no. 4 (2003): 297

be misunderstandings and miscommunications, which, in turn, create mistrust, a poor diagnosis, a wrong prescription, or an inappropriate follow up.[116]

The most crucial challenge facing the organization is to ensure effective communication. In many ways, ineffective communication diminishes individual identity, the validity of the consent, and overlooks obligation. For example, in cases of a terminal disease, there is often less distinction between treatment and research. The information or explanation often comes across to the patient as abstract or merely incomprehensible, which, in turn, leads a patient for a variety of reasons to give uninformed consent.[117] Therefore, proper communication is necessary to provide optimum care throughout healthcare.

Even direct, scientific ways of communication often create confusion. This happens, for instance, when the language of science is used to explain study procedures or when the communication is written in language that the reader struggles to understand.[118] For example, pregnant women face a decision in labor to choose a water birth or an epidural, yet often they remain confused to which choice to make because the consent forms are not written in ways the patient can understand.[119]

The solution to this language challenge is to consciously, systematically, and operationally strive to ensure that all language—both verbal and written—is simple, short, and direct. For complicated and complex cases, such language must not stand-alone. Often in such instances, multimedia serves to facilitate a clearer understanding of the situation, which, in turn, leads to consensus.[120] Practicing Narrative Authority helps to reveal the gap between the perceived and actual services provided. Therefore, the organization should be committed to presenting material to the patient that is understandable and culturally sensitive.

An excellent example of an effective communication practice is the attempt to find consensus. Working toward consensus diminishes the negative effects of power and helps to ensure that each person's autonomy is respected. Even though the physician is expertly trained to diagnose disease, he should still consider that an illness is understood differently by each patient. Hence, the system should ensure practices are in place that allow the physician and patient to see-the-other in the attempt to reach a consensus of understanding and consent to treatment. By exercising Narrative Authority, the system allows physicians to respond to a patient in creative ways that rely heavily on understanding the patient's cultural worldview,

[116] Laurie M. Anderson et al., "Culturally competent healthcare systems," *American Journal of Preventive Medicine* 24, no. 3 (2003): 68–69.

[117] R. A. Kadam, "Informed Consent Process: A Step Further Towards Making It meaningful!", *Perspect Clin Res 8*, no. 3 (2017), 2–4.

[118] R. A. Kadam, "Informed Consent Process: A Step Further Towards Making It Meaningful!", *Perspect Clin Res* 8, no. 3 (2017), 2–6.

[119] Elizabeth Newnham, Lois Mckellar, and Jan Pincombe, "'It's your body, but…' Mixed messages in childbirth education: Findings from a hospital ethnography," *Midwifery* 55 (2017): 53–58.

[120] R. A. Kadam, "Informed Consent Process: A Step Further Towards Making it Meaningful!", *Perspect Clin Res 8*, no. 3 (2017): 6.

her story.[121] For instance, a direct, science-fact way of sharing information with the patient may be counterproductive to someone who is accustomed to hearing good or bad news in a narrative or story-based form. The straightforward, logic-based approach, in this case, would simply be ineffective and inhibit the development of a rich patent-based "therapeutic relationship."[122]

For example, consider a situation when a surgeon attempts to inform a Navajo patient of the details of a proposed bypass surgery. Navajo often do not respond well to direct facts or news that is perceived as negative. While what to the physician appears as a best-case diagnosis, coupled with common and routine procedures, may well be heard by a Navajo patient as a death sentence, who then responds by refusing treatment.[123] In this instance, the surgeon unwittingly undermined the patient's autonomy. He assumed the patient was operating from the same communication perspective and worldview as his, when, in fact, the patient needed to hear the doctor's diagnosis in another way. The direct, science-fact way of sharing his information with the patient may be efficient and logical but was not effective.[124]

Had the surgeon offered his assessment in "positive, ritual language," he could have more effectively communicated his diagnosis and fulfilled his obligation to the patient. Given this scenario, it is important to understand a broader context: "language does not merely describe reality, language shapes reality."[125] Here the provider, procedurally, didn't communicate well with the patient because the power of story was diminished via assumptive thinking. Narrative Authority serves to keep both the organization's and the patient's stories to the fore, through which person-to-person consensus becomes possible.

Communication becomes more problematic when the patient and the provider speak different languages. To address this concern and to improve the clarity and efficacy of communication, healthcare organizations should make an interpreter available. An interpreter can help ensure that both the provider and the patient understand one another, and that the autonomy of the patient is acknowledged and respected. An interpreter helps to ensure both the patient and physician understand the nature and the purpose of the treatment process. An interpreter does more than convey the meaning of the words, he also provides the cultural nuances hidden within the language.[126] Of course, it is not possible to provide all medical materials

---

[121] J. Emilio Carrillo, "Cross-Cultural Primary Care: A Patient-Based Approach," *Annals of Internal Medicine* 130, no. 10 (1999): 832–833.

[122] J. Emilio Carrillo, "Cross-Cultural Primary Care: A Patient-Based Approach," *Annals of Internal Medicine* 130, no. 10 (1999): 830.

[123] Tom L. Beauchamp and James F. Childress, *Principles of biomedical ethics*, 7th ed. (New York: Oxford University Press, 2013), 105–109.

[124] R. A. Kadam, "Informed Consent Process: A Step Further Towards Making It Meaningful!", *Perspect Clin Res* 8, no. 3 (2017), 1–3.

[125] Tom L. Beauchamp and James F. Childress, *Principles of biomedical ethics*, 7th ed. (New York: Oxford University Press, 2013), 109–111.

[126] Laurie M. Anderson et al., "Culturally competent healthcare systems," *American Journal of Preventive Medicine* 24, no. 3 (2003): 69–74.

in every language. In fact, doing so will incur a great cost and loss of resources because no one can predict how many different languages the system will encounter. Therefore, it is cost effective for the system to situationally hire an interpreter and train them to provide better care.[127] However, the interpreter should make sure he or she participates by relying on the principles of Narrative Authority to ensure that the patient's story emerges and fosters partnership between the patient and physician. The interpreter can serve this process by ensuring that both parties understand one another. A skilled interpreter can sense when the communication is out of synch. She can help continue the conversation until she feels the translation has been properly achieved.[128] Narrative Authority guides the interpreter by reminding her that she acts as an intermediary, illuminating details and nuance of the teller to ensure the provider can properly direct the care delivery process. Narrative Authority also offers guidance to ensure that the interpreter does not manipulate or embellish the patient's narrative. When hearing the patient's story or the provider's feedback, the interpreter must ensure that she or he maintains and genuinely expresses the authority of teller—her role is that of narrative-translator.

When there is no interpreter, family members tend to speak on behalf of the patient, which sometimes causes bias and, thus, compromises the patient's autonomy. Therefore, having access to an interpreter also reduces the risk of biased decision-making. The interpreter most often is trained in and knowledgeable of medical terminology and facilitates patient-physician communication. The interpreters are also trained to maintain confidentially; therefore the patient's privacy is protected.[129]

Apart from interpreter service, there are times when negotiation plays a crucial role in effective communication. Negotiation is an important component of communication. This skill helps an individual express her opinions and points of view without dishonoring the other who is involved in the negotiation. For instance, negotiation is important when a patient and physician disagree about a particular medical treatment. In such cases, Narrative Authority drives the negotiation by employing open and respectful listening practices. A negotiation that is driven by Narrative Authority helps the physician to ensure that the patient's preferences are heard and aids the patient to understand the importance of the physician's recommendation. Thus mutual agreement occurs. Consensus is a byproduct of a well-rendered negotiation.[130]

Beyond treatment-option negotiation, negotiation also occurs between care providers; confusion between providers can directly affect care delivery. Confusion and

---

[127] Mary Curry Narayan, "Six Steps toward Cultural Competence: A Clinician's Guide," *Home Health Care Management & Practice* 14, no. 1 (2001): 43–47.

[128] Laurie M. Anderson et al., "Culturally competent healthcare systems," *American Journal of Preventive Medicine* 24, no. 3 (2003): 69–74.

[129] Laurie M. Anderson et al., "Culturally competent healthcare systems," *American Journal of Preventive Medicine* 24, no. 3 (2003): 73–74.

[130] Tineke A. Abma et al., "Inter-Ethics: Towards An Interactive And Interdependent Bioethics," *Bioethics* 24, no. 5 (2010): 253–255.

miscommunication often cause providers to become demotivated and, in turn, provide poor or lackadaisical patient care. A skilled negotiation is a means to solve such problems. This skill can be obtained by applying Narrative Authority. By respecting others and understanding their narratives, it is possible to avoid confusion. The art of negotiation is a very important skill in the exercise of effective, other-centered communication.[131]

Without the approach Narrative Authority provides, it is very difficult to see and understand the other's perspective and to change existing practices. The competency of reflectivity is one of the skills required in today's world. By practicing Narrative Authority, it is possible to maintain an introspective and other-oriented viewpoint that serves to keep present a range of contextual features: social norms, cultural practices, and the nuance of language.[132] Narrative Authority supplies a framework for creating practices that ensure introspection and reflection and serve to have each person see the other as someone with autonomy. Narrative Authority supplies an approach to see the other person, understand her story and concerns, and respond with hospitality, empathy, and respect.[133]

12.4.4 Ensure Education, Ongoing Training and Development

Cultural competence training should be provided for existing and incoming physicians, nurses, and caregivers. Ideally, the training should be offered early in a caregiver's career, before she or he is enveloped in the professional world. The educational system is a natural place to introduce this training and development. Such education and skill building provides opportunities for self-reflection and promotes the beliefs of equity and welcoming the other. Otherwise, the disparities inherent in the system are passed on to the next generation of caregivers.[134] However, often, these efforts also neglect to address the importance of diversifying the makeup of the student body, those who will enter the profession. Having a diverse body of students will help the students and educators become accustomed to diverse narratives. They will learn to share their narratives and create a shared platform where diverse perspectives intersect. The educational system should re-narrate its vision to engender a diverse student body in the school.

However, disparities begin before schooling. Minorities and less privileged populations often cannot find access to a good school. The underprivileged have fewer

[131] Tineke A. Abma et al., "Inter-Ethics: Towards An Interactive And Interdependent Bioethics," *Bioethics* 24, no. 5 (2010): 254–255.

[132] Matthew R. Hunt et al., "The Ethics of Engaged Presence: A Framework for Health Professionals in Humanitarian Assistance and Development Work," *Developing World Bioethics* 14, no. 1 (2012): 53–55.

[133] Liza Ireni Saban, *International Disaster Management Ethics* (Albany: State University of New York Press, 2016): 35–37.

[134] Ursula Småland Goth, Oddhild Bergsli, and Else Marie Johanesen, "Integration of ethnic minorities during group-work for vocational teachers-in-training in health studies," *International Journal of Medical Education* 8 (2017): 31–33.

educational opportunities. Thus, fewer minorities are employed in the healthcare workforce in skilled positions. To make the healthcare system diverse, the system should encourage more diverse student bodies in their schools. The system needs to include health professionals from marginalized backgrounds to create culturally responsive healthcare systems.[135]

Besides bringing diversity to the school, the system should consistently provide diversity training to the students. Skills that promote diversity cannot be attained without necessary training and development. Role-playing, diverse educational resources, and group work help students to think critically and respond to change self-reflectively and in an other-oriented manner. Moral insights and moral questioning arise through experience. Thus, heterogeneous group interaction can help develop future practitioners who see and respect the values of different people, of different stories.[136] Early classroom teaching and interventions help to promote the skills and abilities healthcare providers need to skillfully, respectfully, and confidently handle the nuances of cultural discourse.[137]

Besides providing professional skills on conflict resolution, therapeutic communication, or dress code, the school should arrange clinical training sessions. Culturally sensitive clinical training is a worth-while investment for health professionals because it offers the learner a perspective and "tool" that she can implement and apply in all situations henceforth. Clinical training should include hands-on experience working with a diverse patient population. Clinical training helps professionals to think critically and analyze themselves and others in the throes of this dynamic global arena.[138]

Diversifying the workforce is another way to positively change the existing system to become culturally competitive and responsive. Appointing practitioners, staff, management, and nurses from diverse backgrounds helps create a knowledge- and value-sharing platform in the healthcare system. A minority patient will often feel more comfortable and engaged when she sees a physician from the same or similar ethnic background.[139] On the other hand, an African American patient may feel uncomfortable when her very first interaction in a healthcare system is with staff and care providers who are overwhelmingly white. Having diverse manage-

---

[135] Karline Wilson-Mitchell and Manavi Handa, "Infusing Diversity and Equity Into Clinical Teaching: Training the Trainers," *Journal of Midwifery & Womens Health* 61, no. 6 (2016): 726–728.

[136] Tineke A. Abma et al., "Inter-Ethics: Towards An Interactive And Interdependent Bioethics," *Bioethics* 24, no. 5 (2010): 253–255.

[137] Ursula Småland Goth, Oddhild Bergsli, and Else Marie Johanesen, "Integration of ethnic minorities during group-work for vocational teachers-in-training in health studies," *International Journal of Medical Education* 8 (2017): 31–33.

[138] Karline Wilson-Mitchell and Manavi Handa, "Infusing Diversity and Equity Into Clinical Teaching: Training the Trainers," *Journal of Midwifery & Womens Health* 61, no. 6 (2016): 734–735.

[139] Mary Curry Narayan, "Six Steps toward Cultural Competence: A Clinician's Guide," *Home Health Care Management & Practice* 14, no. 1 (2001): 42–44.

ment teams also will change the mentality of the entire system. These leaders can promote and encourage diversity.[140]

Narrative Authority and the practices it evokes, help to create a personal connection with the *other-as-same*, which, in turn, engenders a culture that feels like one system.[141] To be culturally competent and to encourage diversity, management should integrate the power of Narrative Authority into the mainstream of the decision-making process.

However, training and development efforts must be seen as an ongoing process, founded on the credo that "there is no sufficient level" of cultural competence. The system should reassess its development needs and reflect upon the policies necessary to respond to the growing needs and priorities of a healthcare system operating in a global world. It is now possible for more people to collaborate globally in real time.[142] Therefore, there should be yearly reviews and assessments of these trainings and development efforts to keep them up to date. Management should also keep itself updated and current on best practices. The mainstream of the organization cannot be culturally competent if upper management lacks, discounts the importance of, or fails to practice cultural competency. Trust building activities within the community should also be integrated in the healthcare system.[143] Narrative Authority should sit as the heart of any the healthcare system, at all levels, as a means to build and maintain the trust of and promises made to society.

12.5 Conclusion

To make healthcare culturally competent and culturally responsive, it is necessary for the organization to lead the way and understand its role as healer to and for individuals and society at large. A healthcare organization should be designed in ways that encourage physicians to see patients as individuals, as *others-as-self*, and for patients to see and experience physicians as a caregivers. This humanistic, narrative-driven connection opens a space for dialogue and consensus building. Narrative Authority is a foundational approach and a powerful means to shift and develop practices, processes, and procedures that create an organizational culture where other-oriented caregiving can blossom. Narrative Authority is an approach founded on the belief that humans are radically social and we understand one

[140] Karline Wilson-Mitchell and Manavi Handa, "Infusing Diversity and Equity Into Clinical Teaching: Training the Trainers," *Journal of Midwifery & Womens Health* 61, no. 6 (2016): 726–728.

[141] Clive Baldwin, "Narrative, Ethics and People with Severe Mental Illness," *Australian & New Zealand Journal of Psychiatry* 39, no. 11–12 (2005): 1025–1027.

[142] Thomas L. Friedman, *The World Is Flat: A Brief History of the Twenty-first Century* (Brantford, Ont.: W. Ross MacDonald School Resource Services Library, 2009), 7–8.

[143] Leigh Turner, "From the Local to the Global: Bioethics and the Concept of Culture," *Journal of Medicine and Philosophy* 30, no. 3 (2005): 310–315.

another only through the intersections of our stories, and the coherence and meaning-making that arise through these narrative acts. Embracing this perspective provides a narrative sensitivity that places the caregiver and the *other-in-need-of-care* in an intimate, and, ultimately, loving relationship, founded on the significance that emerges through shared stories. Narrative Authority makes possible the opportunity to construct an ethical healthcare system where practices emerge that are culturally sensitive. Such practices strive to open the space for the other to be seen in and through her story, stripped of stereotype, bias, or prejudice. An exemplary healthcare system is founded on partnerships constructed to promote effective and ethical patient treatment. Narrative Authority further reveals how the organization itself arises from a narrative. This is a narrative that articulates the organization's intent, aspirations, and purpose for its existence. This narrative also reveals core values, normative behaviors, and best practices. It provides an aspirational goal that calls like-minded people to serve—individuals who bring their skills, experiences, and passions to fill required roles to make the intent real, and, in doing so, create an organization that is a narrative-based community. To become an other-oriented organization, this narrative must embed ways to overcome biases, misconceptions, and unquestioned assumptions that blind one to the other and create a darkness where no true healing can occur.

Suggested Readings

The following two texts contain important insights into the key role narrative offers in delivering high-quality, holistic healthcare.

Charon, Rita. 2008. *Narrative Medicine: Honoring the Stories of Illness*. Oxford: Oxford University Press.

Clegg, Gareth, and Colin Robertson. 2017. *Storytelling in Medicine: How Narrative Can Improve Practice*. Boca Raton: CRC Press.

Martha Montello's article on narrative ethics explains the importance of a narrative approach to healthcare.

Martha, Montello. 2014. Narrative Ethics. *Hastings Center Report* 44 (s1): S2–S6.

The critical analysis for the concept of Narrative Authority that is presented in this chapter is influenced by Frank Lehner, who describes the workplace as a social structure that is best understood as an identity-based coalition bound together in a common narrative engaged in routine innovation.

Lehner, Frank. 2012. *The Ingenious Workplace: An Antidote To Common Workplace Maladies*. Pittsburgh: Naridus, LLC.

References

Abma, Tineke A., Vivianne E. Baur, Bert Molewijk, and Guy A.M. Widdershoven. 2010. Inter-Ethics: Towards an Interactive and Interdependent Bioethics. *Bioethics* 24 (5): 242–255.

Anderson, Laurie M., Susan C. Scrimshaw, Mindy T. Fullilove, Jonathan E. Fielding, and Jacques Normand. 2003. Culturally Competent Healthcare Systems. *American Journal of Preventive Medicine* 24 (3): 68–79.

Baldwin, Clive. 2005. Narrative, Ethics and People with Severe Mental Illness. *Australian and New Zealand Journal of Psychiatry* 39 (11–12): 1022–1029.

Blackhall, Leslie J., Gelya Frank, Sheila T. Murphy, Vicki Michel, Joycelynne M. Palmer, and Stanley P. Azen. 1999. Ethnicity and Attitudes Towards Life Sustaining Technology. *Social Science & Medicine* 48 (12): 1779–1789.

Beauchamp, Tom L., and James F. Childress. 2009. *Principles of Biomedical Ethics*. New York: Oxford University Press. (in English).

Berlin, E.A., and W.C. Fowkes Jr. 1983. A Teaching Framework for Cross-Cultural Health Care. Application in Family Practice. *The Western Journal of Medicine* 139 (6): 934–938. (in English).

Betancourt, J.R., A.R. Green, J.E. Carrillo, and O. Ananeh-Firempong 2nd. 2003. Defining Cultural Competence: A Practical Framework for Addressing Racial/Ethnic Disparities in Health and Health Care. *Public Health Reports* 118 (4): 293–302. (in English).

Burgess, Diana Jill, Megan Crowley-Matoka, Sean Phelan, John F. Dovidio, Robert Kerns, Craig Roth, Somnath Saha, and Michelle van Ryn. 2008. Patient Race and Physicians' Decisions to Prescribe Opioids for Chronic Low Back Pain. *Social Science & Medicine* 67 (11): 1852–1860.

Buetow, S.A. 2005. To Care Is to Coprovide. *The Annals of Family Medicine* 3 (6): 553–555.

Carrillo, J.Emilio. 1999. Cross-Cultural Primary Care: A Patient-Based Approach. *Annals of Internal Medicine* 130 (10): 829–834.

Charon, Rita. 2001. Narrative Medicine. *JAMA* 286 (15): 1897.

———. 2008. *Narrative Medicine: Honoring the Stories of Illness*. Oxford: Oxford University Press.

Dahrouge, Simone, William Hogg, Meltem Tuna, Grant Russell, Rose Anne Devlin, Peter Tugwell, and Elisabeth Kristjansson. 2010. An Evaluation of Gender Equity in Different Models of Primary Care Practices in Ontario. *BMC Public Health* 10 (1): 1–12.

Friedman, Thomas L. 2009. *The World Is Flat: A Brief History of the Twenty-first Century*. Brantford: W. Ross MacDonald School Resource Services Library.

Globalization of Healthcare. *Global Advances in Health and Medicine* 1(2) (05/01 2012).

Graham, Iain, and Linda Norman. 2008. A Reflective Discussion: Questions About Globalization and Multiculturalism in Nursing as Revealed During a Student/Staff Exchange Programme. *International Journal of Nursing Practice* 14 (3): 189–194.

Goth, Ursula Småland, Oddhild Bergsli, and Else Marie Johanesen. 2017. Integration of Ethnic Minorities During Group-Work for Vocational Teachers-in-Training in Health Studies. *International Journal of Medical Education* 8: 31–36.

Hasnain-Wynia, Romana, Raymond Kang, Mary Beth Landrum, Christine Vogeli, David W. Baker, and Joel S. Weissman. 2010. Racial and Ethnic Disparities Within and Between Hospitals for Inpatient Quality of Care: An Examination of Patient-Level Hospital Quality Alliance Measures. *Journal of Health Care for the Poor and Underserved* 21 (2): 629–648.

Harrell, Regina. 2013, November 17. Why a Patient's Story Matters More Than a Computer Checklist. NPR. Accessed 23 June 2018. https://www.npr.org/sections/health-shots/2013/11/17/242366259/why-a-patient-s-story-matters-more-than-a-computer-checklist

Hellsten, Sirkku K. 2008. Global Bioethics: Utopia or Reality? *Developing World Bioethics* 8 (2): 70–81.

Jones, A.H. 1999. Narrative Based Medicine: Narrative in Medical Ethics. *BMJ* 318 (7178): 253–256.

Kadam, R.A. 2017. Informed Consent Process: A Step Further Towards Making Meaningful! *Perspectives in Clinical Research* 8: 1–9. (in English).

Kahissay, Mesfin H., Teferi G. Fenta, and Heather Boon. 2017. Beliefs and Perception of Ill-Health Causation: A Socio-Cultural Qualitative Study in Rural North-Eastern Ethiopia. *BMC Public Health* 17 (1): 1–10.

Kim, Jin Un, Obinna Oleribe, Ramou Njie, and Simon Taylor-Robinson. 2017. A Time for New North–South Relationships in Global Health. *International Journal of General Medicine* 10: 401–408.

Lee, K. 2000. For Debate. The Impact of Globalization on Public Health: Implications for the UK Faculty of Public Health Medicine. *Journal of Public Health* 22 (3): 253–262.

Lehner, Frank. 2012. *The Ingenious Workplace: An Antidote To Common Workplace Maladies*. Pittsburgh: Naridus, LLC.

Macklin, Ruth. 2004. Double Standards in Medical Research in Developing Countries. SCRIPT-ed, 639–640.

Mars, Godelief M.J., Gertrudis I.J.M. Kempen, Guy A.M. Widdershoven, Peter P.M. Janssen, and Jacques T.M. Van Eijk. 2008. Conceptualizing Autonomy in the Context of Chronic Physical Illness: Relating Philosophical Theories to Social Scientific Perspectives. *Health: An Interdisciplinary Journal for the Social Study of Health, Illness and Medicine* 12 (3): 333–348.

Mbugua, Karori. 2012. Respect for Cultural Diversity and the Empirical Turn in Bioethics: A Plea for Caution. *Journal of Medical Ethics and History of Medicine* 5: 2–5.

Minorities expected to be majority in 2050. CNN. Accessed 28 Nov 2017. http://www.cnn.com/2008/US/08/13/census.minorities/index.html?_s=PM%3AUS

Morris, David B. 2001. Narrative, Ethics, and Pain: Thinking with Stories. *Narrative* 1: 9.

Montello, Martha. 2014. Narrative Ethics. *Hastings Center Report* 44 (S1): S2–S6.

Narayan, Mary Curry. 2001. Six Steps Toward Cultural Competence: A Clinician's Guide. *Home Health Care Management & Practice* 14 (1): 40–48.

Newnham, Elizabeth, Lois Mckellar, and Jan Pincombe. 2017. Its Your Body, but…' Mixed Messages in Childbirth Education: Findings from a Hospital Ethnography. *Midwifery* 55: 53–59.

Newton, Adam Zachary. 1997. *Narrative Ethics*. Cambridge, MA: Harvard University Press.

Oflynn, N., and S. Staniszewska. 2012. Improving the Experience of Care for People Using NHS Services: Summary of NICE Guidance. *BMJ* 344 (Mar16 1): D6422.

Padela, Aasim I., Aisha Y. Malik, Farr Curlin, and Raymond De Vries. 2014. [Re]Considering Respect for Persons in a Globalizing World. *Developing World Bioethics* 15 (2): 98–106.

Robertson, Colin, and Gareth Clegg. 2017. *Storytelling in Medicine: How Narrative Can Improve Practice*. Boca Raton: CRC Press/Taylor & Francis Group.

Rovelli, Carlo, Simon Carnell, and Erica Segre. 2016. *Seven Brief Lessons on Physics*. New York: Riverhead Books.

Saikia, Nandita, Moradhvaj, and Jayanta Kumar Bora. 2016. Gender Difference in Health-Care Expenditure: Evidence from India Human Development Survey. *PLoS One* 11 (7): e0158332.

Saha, Somnath, Mary Catherine Beach, and Lisa A. Cooper. 2008. Patient Centeredness, Cultural Competence and Healthcare Quality. *Journal of the National Medical Association* 100 (11): 1275–1285.

Segouin, C. 2005. Globalization in Health Care: Is International Standardization of Quality a Step Toward Outsourcing? *International Journal for Quality in Health Care* 17 (4): 277–279.

Sue, Derald Wing. 2001. Multidimensional Facets of Cultural Competence. *The Counseling Psychologist* 29 (6): 790–821.

Tai, M.C.-T. 2001. Developing a Culturally Relevant Bioethics for Asian People. *Journal of Medical Ethics* 27 (1): 51–54.

Turner, Leigh. 2003. Bioethics in a Multicultural World: Medicine and Morality in Pluralistic Settings. *Health Care Analysis* 11 (2): 99–117.

———. 2005. From the Local to the Global: Bioethics and the Concept of Culture. *Journal of Medicine and Philosophy* 30 (3): 305–320.

Veatch, Robert M. 2009. *Patient, Heal Thyself: How the New Medicine Puts the Patient in Charge*. Oxford: Oxford University Press.

Washington, Donna L., Jacqueline Bowles, Somnath Saha, Carol R. Horowitz, Sandra Moody-Ayers, Arleen F. Brown, Valerie E. Stone, and Lisa A. Cooper. 2008. Transforming Clinical Practice to Eliminate Racial–Ethnic Disparities in Healthcare. *Journal of General Internal Medicine* 23 (5): 685–691.

Why US Health Care Should Think Globally. 2016. *The AMA Journal of Ethic* 18 (7): 736–742.

Wilson-Mitchell, Karline, and Manavi Handa. 2016. Infusing Diversity and Equity into Clinical Teaching: Training the Trainers. *Journal of Midwifery & Womens Health* 61 (6): 726–736.

Zomorodi, Meg, and Barbara Jo Foley. 2009. The Nature of Advocacy vs. Paternalism in Nursing: Clarifying the 'Thin Line'. *Journal of Advanced Nursing* 65 (8): 1746–1752.

Chapter 13
The Authority of the Ethics Consultant

Kenneth Kipnis

Abstract It is no longer news that philosophers specializing in ethics have begun to serve as consultants, in hospitals and courtrooms for example. What is not now as clear as it should be is the source of the authority these new specialists claim on matters ethical. What is it they can bring to puzzlement? Should laymen defer to their superior judgment and, if so, why? At trial, expert witnesses must establish their qualifications in court before they will be allowed to weigh in on legal controversies. Using a single court case as a paradigm, I will reflect on the intellectual foundations of the work of the expert ethics witness and argue for the clarity that ethics expertise can bring to consequential perplexity.

Keywords Pluralism · John Rawls · Ethics consultation · Ethics practitioner · Ethicist · Christine Korsgaard · Normativity · Authority (of position and expertise) · G. A. Cohen · Expert testimony · American Psychological Association · Conflict of interest · Reflective endorsement · Practical identity · Liam O'Flaherty · Ethical theory · Professional ethics

13.1 The Samoan

I have lived in Honolulu for more than 35 years, teaching "ethics in health care" in many different settings. Unlike most places, Hawaii has no majority population. We are all minorities here. Caucasians represent only about a third of the population. The rest are persons of Japanese, Chinese, Filipino, Hawaiian, Samoan, Korean, Puerto Rican, Native American, Portuguese, or African origin and many others.

My medical ethics students have taught me to be cautious about assuming consensus, for the moralities they bring with them into medical education are strikingly diverse. At the beginning of my course, I regularly tell them about a 25-year old

K. Kipnis (✉)
Professor Emeritus of Philosophy, University of Hawaii, Honolulu, HI, USA

J. Gielen (ed.), *Dealing with Bioethical Issues in a Globalized World*, Advancing Global Bioethics 14, https://doi.org/10.1007/978-3-030-30432-4_13

Samoan who had a tumor in his leg. It had metastasized and death could not be prevented. Most of us who live in Hawaii know that Samoan families are tightly-knit, that sons and daughters acknowledge a far greater deference to parental authority than Westerners do, and that their families crowd into hospital rooms when a loved one is sick. So the relatives were gathered at the hospital bed for a meeting with the doctor who was recommending amputation, to prolong the young man's life for perhaps a few additional months. The father, who was also the chief, approved the plan and the young man nodded consent. But later on, alone now in preop, the patient confides to the surgeon that he does not want to live out his last weeks as an amputee. Even more, he does not want his father, the chief, to know he has questioned his judgment. For the young man, the social consequences of such disclosure would be worse than amputation.

I ask my students, "What should a good doctor do?" For the Samoans in my class, the question is a no-brainer. Take the leg off. Why are we talking about this? Caucasians find some way to backtrack, telling the father a story if necessary, but absolutely not doing the surgery. Asians and Pacific islanders are split, some siding with the Caucasians, but others honoring the chief's decision either because the young man should be treated under Samoan rules or out of a sense of filial respect. It gradually becomes clear to the class that there is no consensus; that – collectively – they do not know what a good doctor should do. This acknowledgment is the beginning of wisdom.

I recount this story here because what is true of Hawaii is becoming true in most places. Pluralism, as John Rawls tells us, has become a permanent feature of the human condition, not likely to fade away. The representative peoples in my medical ethics class need to be able to bracket somehow the personal moralities they bring with them into the profession. They need to open themselves to the possibility that, in becoming physicians, they may have to learn new and different ways of understanding their obligations.

13.2 "Just Who Do You Think You Are?"[1]

It is no longer news that academic philosophers are sallying beyond their campuses to do ethics consultation. For some, emigration has become permanent. These expatriates have severed ties with former professional colleagues and diminished their identification with the discipline of philosophy. Others, myself included, retain our day jobs. We continue to cover traditional courses while carving out unconventional niches extending across departmental and academic frontiers. Unlike the expatriates, we aspire to maintain dual citizenship, becoming multi–lingual as needed and managing the divided attentions and loyalties that characterize such a career. Ethics practitioners, ethicists, ethics consultants (terminology proliferates) now work in hospitals, research organizations, corporations, government agencies, legislatures,

[1] The question is from Korsgaard, *The Sources of Normativity* (1996), p. 101.

professional associations, the courts, and with individual clients. Although I have done work in all of these settings, this essay focuses on my service as an expert witness in a single ethics–related court case.

Exactly what is meant by the "authority" of the ethics consultant? What is it that can endow certain ethical judgments with the power to impose themselves upon our practical deliberations? Though her account has drawn criticism, Christine Korsgaard calls this feature of morality "normativity," a kind of ethical *gravitas*. Some account of normativity would seem to be critical to anyone claiming ethical authority. What authority (if any) can ethics consultants summon for the judgments that constitute their professional work?

It is commonplace to distinguish between two conceptions of "authority." The first focuses on the social role occupied by one who is "in authority": Call this the Authority of Position. One can usefully think of relationships, complex activities, societies and organizations as constructed out of "normatively rich" social roles, slots that carry with them formal obligations and permissions.

In a university classroom, professors have the positional authority to assign readings and issue grades. An ethics professor might believe (as I once did) that with a lengthy specialization in ethics, a doctoral degree, and the blessing of some accredited university, he or she is somehow empowered to pronounce authoritatively upon ethical questions. One utters the words "In my professional judgment …" and follows it with some *ex cathedra* utterance to be taken to heart. But it would be a mistake to suppose that ethicists, ethics consultants, professors of philosophy, or academic philosophers have some God–like power to create moral obligations merely by solemnly pronouncing them. Lest there be any misunderstanding, I claim no such power.

The second conception of authority – call this the Authority of Expertise – points to the special knowledge possessed by the expert. One can be an authority on high–energy physics, bariatric surgery, the Irish Civil War, Mafia morality, and so on. In each case, authority consists in the mastery of some sphere of expertise. So what exactly is the knowledge that an ethics consultant possesses that could warrant the ethical judgments that he or she might offer? How could this happen and could the work product be intellectually sound?

Today, there are live questions about what ethics consultants are, and what we could be: how we might someday conceive ourselves as professionals. The ambiguous use of the term "conceive" in the last sentence is revealing. The term can refer to an act of intellectual imagination: An author might "conceive" the plot of a novel. Or the term can reference the coming into being of some new entity: A generation of college students might conceive a novel political movement. In talking about "conceiving oneself" as an ethics consultant, both meanings can be in play. Indeed, a professional typically comes into being during the course of systematic education. He or she begins to assume a new practical identity. Similarly, a new profession (a new social role) can emerge from the collective creative activities of reflective practitioners. Social roles are historically situated works of our individual and collective imaginations – human artifacts – and, at the same time, they are palpable social entities out of which we construct relationships and organizations. Like all human artifacts, social roles can be crafted with wisdom or foolishness. Responsible communities must stand ready to assess and reconfigure these creations when the occasion calls for it. A new social role – the professional ethics consultant – may be emerging.

So what is an ethics consultant? What are the proper dimensions of the moral authority that the ethics consultant wields? In what has been a multifaceted career, I have had to grapple with this question in several different settings. When I was starting out as an undergraduate philosophy student, the conventional wisdom was that philosophers had nothing to say about substantive ethical problems. Since then, I have had to revise my thinking on many occasions. Indeed, to be a philosopher is to advance intellectual positions in the face of severe criticism. If you weren't ready to change your mind, you had made an error in career choice. Over time, as I groped toward an understanding of what an ethics consultant could be, I learned that there are few critics more challenging than the lawyers whose job it was to demolish expert witnesses. Much of this paper emerged from being grilled at deposition by a well-prepared attorney.

13.3 The Case

Not very long ago I was contacted by a law firm with a request that I consider working for them as an expert witness. It was not the first time I had done such work and the facts of this case were not complicated. The plaintiff (the party filing a complaint to the court) was a man whose life had taken a tragic turn. He had retained a psychologist to help him get back on his feet and, following improvement, their professional relationship had effectively come to an end and their association evolved into something like a friendship. The therapist stayed at the former patient's house, the two men vacationed together, and the psychologist became a valued source of advice, particularly on financial matters. Eventually the therapist arranged the transfer of substantial amounts of the former client's wealth into a seemingly lucrative investment. But the "investment" turned out to be a swindle. Any profits that the scheme's longer–term "investors" might take would be paid out of sums that new investors were transferring to the fund manager who was enjoying significant wealth. The psychologist was one of several "finders" receiving payment for each mark he recruited. Scams like these are called "Ponzi schemes", after Charles Ponzi whose victims famously lost millions before he went to jail in 1920.

This case was a civil lawsuit that turned, in part, on whether the psychologist had violated the norms of his profession. Though it was not something I ever discussed, I had assumed that the duty of some insurance company to pay damages to the client would turn on whether the losses were covered by the psychologist's professional insurance policy. And that would depend on whether the injuries had occurred in the context of the psychologist's professional work. If the misconduct had occurred outside the ambit of the professional relationship – say he ran over his patient in the clinic's parking lot – the insurance company would be off the hook. The key question was whether the psychologist's behavior violated the ethical duties of his profession, duties that the law would recognize.

As a matter of law, suits such as this one required the plaintiff's attorney to prove three elements in court. First, was the psychologist subject to some duty recognized by law? The norms of professional ethics qualified as such, but identifying them and explaining their scope and justification to a judge and jury are rarely easy. Were there applicable ethical standards, and, if so, were they mere technicalities, easily ignored, or were they rules that reasonable people would want their therapists to honor? This area was my responsibility. The second element to be proved in court was that the defendant had failed to conform to the duty established under the first element. I had no role in establishing those facts. And the third element was that the plaintiff had suffered compensable injuries caused by the violation of the duty. The evidence for this element also fell to others. My concerns were narrowly with the first element: What pertinent professional duties were applicable to the psychologist? This was, without doubt, a question of ethics and, under the circumstances, a great deal turned on how it was answered.

As a first cut, I would try to achieve a broad familiarity with the ethical standards applicable to psychology. While it was useful, as background, to have chatted informally with experienced practitioners, The American Psychological Association's (APA) ***Ethical Principles of Psychologists and Code of Conduct*** (the Ethics Code: American Psychological Association 2010), carried much more weight. Although lawyers and journalists speak of conflicts of interest, and doctors sometimes refer to "double agent" problems, the parallel idea of a "multiple relationship" was central to the ethical thinking of psychologists, both in the American Psychological Association's code and the profession's literature. The key language is in section 3.05 of the code:

> **Multiple Relationships**
>
> (a) A multiple relationship occurs when a psychologist is in a professional role with a person and … at the same time is in another role with the same person …
>
> A psychologist refrains from entering into a multiple relationship if the multiple relationship could reasonably be expected to impair the psychologist's objectivity, competence or effectiveness in performing his or her functions as a psychologist, or otherwise risks exploitation or harm to the person with whom the professional relationship exists.

If one took the plain meaning of this language to be definitive of the professional's distinctive responsibility, then a surface reading of the code would support the psychologist. It would be argued that, since the professional relationship between the therapist and the client had terminated, there never was a time when the psychologist simultaneously occupied the roles of financial advisor and therapist. Since the two relationships were not in place "at the same time" (as required by the Ethics Code), the psychologist's behavior, however dishonorable, was not a violation of professional ethics. Accordingly, the injuries done to the former client did not occur within the context of the therapist's professional work and the therapist's insurance carrier would not have to compensate the former patient for the losses caused by the psychologist. My lawyer and his client would lose.

There is more to this discussion below. But before moving on, it would be useful to come back to the work of Christine Korsgaard.

13.4 Korsgaard's Account of Normativity

Korsgaard's account of ethics begins with the observations that ethical standards "make *claims* on us; they command, oblige, recommend, or guide" (Korsgaard 1996, p. 2) – and that honoring the demands of morality can be burdensome and costly. So, Korsgaard asks, why be moral? What is it that makes us susceptible to morality's influence even when the costs are high? (Korsgaard 1996, p. 17)

Korsgaard's moral psychology focuses on the human mind's characteristic capacity for self–reflection. We may be unique in our ability to stand back from our desires, inclinations, drives, motivations, etc., and ask: Do I have a sound reason to act? Confronted with an ethically problematic situation, it is open to us to reflect. The identification and appropriation of a reason to act is, Korsgaard says, the hallmark of deliberative success. It is this capacity for "reflective endorsement" that sets us apart from creatures driven by passions. To embrace and act upon a reason – a "reason" being the distinctive product of practical deliberation – is to defer to the authority of reflection. Autonomy, as Korsgaard puts it, is "commanding yourself to do what you think it would be a good idea to do."

But why defer to the commands of reason when doing so is burdensome? Here Korsgaard adds a new and distinctive element to a broadly Kantian account of morality. What you think you have a good reason to do, Korsgaard says, "depends on who you think you are." (Korsgaard 1996, p. 107) In a section of her book entitled "The Solution" (3.3.1), Korsgaard writes:

> When you deliberate, it is as if there were something over and above all of your desires, something which is *you*, and which *chooses* which desire to act on.

The "something" that is "over and above all of your desires" is, according to Korsgaard, a "description under which you value yourself … a conception of your practical identity." She reminds us of the familiar "I couldn't live with myself if I did that."

According to Korsgaard, the ways we conceive ourselves can give rise to "unconditional obligations." (Korsgaard 1996, p. 102) To embrace a practical identity is to identify with a principle, a "way of choosing" (Korsgaard 1996, p. 100); it is to become "a law unto yourself" (Korsgaard 1996, 104), an autonomous agent. To conceive yourself as a friend, a comrade–in–arms, a teacher, a lawyer, a citizen, a doctor, a son, a mother, a psychologist, a child of God, a student, an ethicist, or an Irish patriot, is to endorse reflectively the norms that somehow flow from these practical identities.

> When an action cannot be performed without loss of some fundamental part of one's identity, and an agent could just as well be dead, the obligation not to do it is unconditional and complete. If reasons arise from reflective endorsement, then obligation arises from reflective *rejection*. (Korsgaard 1996, p. 102)

I think the point Korsgaard is making here is sound. To betray one's fundamental obligations can be to cast oneself into the abyss. John Ford's 1935 movie, *The Informer*, based on Liam O'Flaherty's novel (O'Flaherty 1961), illustrates the possibility. Gypo Nolan is a burly, dull–witted Irishman, actively sympathetic to the Irish insurgency against the British in the 1920s, but also eager to leave an Ireland that has fallen on terrible times. He sees one of his political comrades on a wanted poster, enters a police station and, to collect the reward, tells the officers where they can apprehend him. Later on, at his rebel–conducted trial, Gypo's self–conception is undone as he confronts, through an alcoholic haze, his betrayal of a comrade–in–arms. In Liam O'Flaherty's words, he became a "phantom of a human soul stripped naked of the covering of civilization … horror–stricken, without help, without hope of mercy" (O'Flaherty 1961, p. 133). In Korsgaardian terms, he has demolished his practical identity as an Irish patriot and friend. Condemned as an informer, Gypo's death, at the hands of his former comrades, comes quickly.

For Korsgaard, practical identity is the bedrock upon which normativity rests. Her account gives us an explanation for firm commitments to ethical principles – rules that are somehow built into our practical identities. From the perspective of practical ethics, her work is, I believe, an important contribution.

13.5 Cohen's 'Mafioso' Objection

In the first of the five critical essays included in *Sources of Normativity*, G.A. Cohen asks us to consider an idealized "Mafioso" whose practical identity requires the performance of contemptible misdeeds (Cohen 1996, p. 183). The Mafia norm of *omertá* can require a *capo* to order the execution of a soldier who is collaborating with the police.

If, following Korsgaard's approach, Cohen's Mafioso were to reflect upon his practical identity and its normative implications, if the description under which he valued himself were "man of honor" (as the Mafia understood that social role), his deliberations would issue in a reasoned decision to conform to the code defining his practical identity. He will order the death of the collaborator. I assume here that, as with Cohen, most of us would judge the gangland killing to be an ethical violation.

Cohen considers the possibility that Korsgaard has given us merely an account of "the experience or phenomenology of obligation: not its ground or authenticating source." If so, this would rule out reading her work as an essay in normative ethics. The "justification" for normativity that her theory promises cannot be delivered using the tools she has assembled. As Cohen's example seems to show, the bare fact of successful reflection on the normative implications of one's practical identity cannot be the foundation of moral rightness. Perhaps something must be added to the mix to make a Korsgaardian obligation into an ethical one. And, if that were done, why the Mafioso lacks that essential ingredient would need to be explained.

13.6 From Practical Identity to Social Role

It is a critical step in Korsgaard's analysis that the norms that can compel us to act flow somehow from a practical identity. But her text is not very clear how ethical obligations are generated in this way. As she conceives it, one's "practical identity" is a part of one's moral psychology, and while her discussion may suffice to explain why one might fervently believe one has a good reason to act in some way, it fails to explain why acting in that way would be *ethically* required. That, I think, is the thrust of Cohen's objection. Fortunately, a foundation for such ethical obligations can, I believe, be located close by.

Let us turn our attention away from moral psychology, and consider that the concept of practical identity, upon which Korsgaard places so much theoretical weight, can be supplemented by the concrete ways in which our practical identities make their appearance in the public sphere: as persons who occupy discrete and familiar social roles. For Korsgaard, a "practical identity" is a *psychological entity* answering the question How do I conceive myself? In contrast, a "social role" is a *societal entity* answering the question What parts do I play in relationships and organizations? While it is possible for the two to diverge – think of Don Quixote foolishly conceiving himself as a "knight-errant" – they are characteristically two sides of the same coin. One of my practical identities is "philosophy professor" and, in fact, I have occupied such a position.

The normative dimension of social roles is particularly evident in the professions, where the collective attention of practitioners can focus on the ethical principles and ideals that are built into the role itself. It can be puzzling when one steeped in philosophical ethics discovers that the special obligations attaching to professional roles can be in conflict with obligations that are often believed to be applicable to everyone. The contrast between the demands of ordinary morality and the normative implications of professional responsibility can be striking. Although most people accept that no one should assist wrongdoers to evade the just consequences of their misdeeds – driving the getaway car after a bank robbery, for example – defense attorneys may take credit for securing a felon's acquittal. Accordingly if it is permissible to be a defense attorney, lawyers are subject to different ethical standards. Confidentiality offers striking contrasts between professional obligations and everyday norms.

As an occupation develops into a profession, practitioners begin to conceive themselves as colleagues who have, collectively, taken on a responsibility to attend appropriately to one or more important social goods. Good doctors, for instance, play a determinate part in a broad and complex social undertaking aimed at healing the ill. The provisions in a profession's code of ethics are norms that, provided they are generally honored by practitioners, will ensure that the sphere of social concern delegated to and appropriated by a profession as its distinctive collective responsibility, receives the care and attention it is due. Our health care and legal systems – both critically important social institutions – would not work as well as we expect them to if practitioners failed to care deeply about applicable professional standards. The norms that shape professional duties are attached publicly to codes that authoritatively promulgate standards that are themselves the product of collective deliberation.

Philosophers are not the only ones who think seriously and systematically about ethics. The members of professions do so as well. When a profession is in good order, experienced professionals will deliberate about the ethical dimensions of professional practice even as novices internalize the standards together with the reasonings that ground them. And practitioners who egregiously breach the standards – even to serve socially beneficent purposes – may lose their licenses to practice. As one studies the literatures of professional ethics, one finds arguments advanced for role–specific obligations, including, on occasion, obligations that diverge strikingly from "ordinary morality." What can ground these obligations can be a collective social responsibility to provide care and attention to some critical arena of societal concern. Provisions in ethical codes precipitate out of the profession's responsibility for some overriding constituting task.

I want to suggest that the distinction between the normativity of the Mafia's commitment to *omertá*, and the normativity of the legal and medical profession's commitments to confidentiality, becomes evident when we assess the social organizations in question. If we think through the issues, we will likely be led to the judgment that, in the interest of social justice, societies like ours can require their lawyers to be honorable in determinate ways. Likewise, they require doctors to comply with norms that permit health care systems to function well, securing a community's trust and physical well–being as well as the arts and sciences of medicine will allow. When the institutions we rely upon for justice and health care are functioning satisfactorily, we are all more likely to be better off, at least in those respects.

From the point of view of ethical theory, there is a consequentialist element to this picture; professions that are in good order further and secure discrete but significant social values. And there is a deontological element as well, in the way a novice acquires distinctive social obligations during the course of entry into a profession, coming to share responsibility for its constituting task.

In striking contrast, it is not now a benefit to society that our mobsters honor the venerable precepts of *omertá*. In the near anarchy of mid nineteenth century Sicily, it might have been plausibly argued that the Mafia was necessary as a "mutual protective association." Regardless of the social purposes that Mafia families may have served then, it now seems that communities no longer require such organizations, and should no longer be extorted into paying for them.

Viewed in this narrow way, the complaint that will stand against the Mafioso is not that he is a bad person, nor that he violates commonly accepted norms, nor that his behavior conflicts with the practical identity that he has, nor a practical identity that he should have, though all of these complaints may be well–grounded in the end: It is rather that he works within an organization that could not exist within a just social order. There are better ways of securing the societal goods formerly "entrusted" to Mafia families.

When we shift our attention back and forth, between the personal side (our practical identities) and the public side (our social roles and the institutions and relationships of which they are a part), we have, I think, a solution to Korsgaard's problem, at least in cases involving professional obligations. While the experience of obligation might be grounded upon practical identity, at least some *ethical* obligations are grounded when practical identity is analytically linked to a normatively rich social role, and the role in question is linked to a social relationship or organization that is critical to the well–being of a social order.

13.7 The Psychologist's Culpability

We can now return to the question of the psychologist's professional culpability and the role I played as an expert witness. In keeping with the previous discussion, note that "psychologist" names a normatively rich social role, and that the craft is one that can contribute to human well-being. Sound professional codes mandate practices that enable a profession to provide due care and attention to the values that inform the profession's work.

It has become my practice to draw upon two sources in assembling testimony as an expert ethics witness: (1) the ethical provisions in the consensus documents, and (2) the associated arguments drawn from the secondary literature. My deposition lasted for about 6 h. Here I can only sketch some main lines of what I said.

For many experts, pertinent obligations are thought to reside exclusively in the black letters of the document. In contrast, it seemed to me that the norms specified in authoritative documents were not themselves authoritative. They were not like divine commands, worthy of endorsement merely because they had been inscribed on the profession's Holy Tablets. They were in the code just because they had been, at the time they were drafted, taken to be sound ethical precepts. In that sense, the code was not the source of the professional's special obligations. Rather, the Ethics Code was the profession's public acknowledgement of the scope of its obligations. Following Korsgaard, the sources of the profession's normativity were to be found in the good reasons for its ethical provisions.

Accordingly, it could not settle a question of professional ethics that the American Psychological Association's Code said that, (with one exception to be discussed below) multiple relationships commencing after the termination of treatment were permissible. What interested me were the arguments from the secondary literature, those that could be brought to bear on the limits of permissible multiple relationships, especially those that might commence after the termination of treatment.

This deep and broad familiarity with arguments and reasons is at the heart of the ethics consultant's distinctive expertise. It is a species of the Authority of Expertise. Ideally the ethics expert will be a curator of reasons: well–versed in the ethics literature of the profession in question and with parallel discussions from other professions. The question on the table will be: What are the good arguments for configuring professional obligations in one way rather than another?

In a courtroom, it will not be enough merely to have mastered those arguments. One will also have to have the ability to represent them in ways that are clear enough for a judge and jury to follow. (Also important: opposing counsel must be able to appreciate that both the judge and the jury will be able to follow the arguments. The good expert witness is, above all, a good teacher.)

As I continued to read and re–read the code, a puzzling element came into focus. Far removed from the "multiple relationship" language was a discussion of sexual relationships with former clients. In a section of the Ethics Code cryptically entitled "Therapy," there were the following provisions:

> **10.08 Sexual Intimacies with Former Therapy Clients/Patients**
> (a) Psychologists do not engage in sexual intimacies with former clients/patients for at least two years after cessation or termination of therapy.
> (b) Psychologists do not engage in sexual intimacies with former clients/patients even after a two–year interval except in the most unusual circumstances. Psychologists who engage in such activity after the two years following cessation or termination of therapy and of having no sexual contact with the former client/patient bear the burden of demonstrating that there has been no exploitation, … (See also Standard 3.05, Multiple Relationships.)

The parenthetical reference to Standard 3.05 (on "Multiple Relationships") suggests that the authors of the code were aware that a sexual or romantic relationship would qualify as a multiple relationship as discussed earlier. So why, one might wonder, was the issue of sexual intimacy with former clients not subsumed under "Multiple Relationships"? Further, why did the prohibition on multiple relationships expire when therapy terminated, even as the prohibition on sexual relationships survived for at least 2 years following termination? What was it that justified the difference?

Having done work on professional ethics in several fields, I was aware that professional obligations often survived the termination of the fiduciary relationship: for example, the duties to preserve confidences and maintain accurate and complete records. But there were also cases that highlighted other problems created by successive relationships. For example, a lawyer who had once done work for a businessman would not be permitted to represent, later on, the businessman's wife in her divorce with the former client. Consider that if the lawyer agreed to represent the wife, he would have to stand ready to use all of his resources against the husband. Such is the paramount duty of zealous advocacy. And yet, because of confidentiality, he would be prohibited from using what he had learned about the businessman while he was acting as the lawyer for that former client. If financial information earlier obtained from the husband were potentially useful in the wife's case, he would have had conflicting obligations both to use that information on behalf of his second client and not to use that same information out of an obligation to his first client. For this reason legal ethics prohibits attorneys from entering into successive representations where such conflicts of obligation are possible.

The authors of the APA Ethics Code seem to have been aware of such risks. In delimiting the prohibition against sexual intimacy, the Ethics Code permits sexual intimacy with a former client after a two–year post–termination period, provided that there have been no sexual relations beforehand and provided that the psychologist can demonstrate that there has been no exploitation of the client.

But notice that a restrictive, formalistic interpretation of the code would still support the defense. While sexual intimacy was the sole exception to the rule prohibiting only simultaneous multiple relationships, the psychologist's complicity in the swindle did not remotely qualify as a sexual relationship, the one designated exception. Strictly construed, it didn't matter that the secondary relationship (financial manager) did more than merely risk exploitation and harm: it richly visited both upon the former patient. My client would still lose.

As it happened, in a section entitled Introduction and Applicability, the APA Ethics Code explicitly discourages restrictive, formalistic interpretations.

> In applying the Ethics Code to their professional work, psychologists may consider other materials and guidelines that have been adopted or endorsed by scientific and professional psychological organizations ... as well as consult with others within the field. [APA Code: Introduction and Applicability]

I took this language to be a formal encouragement of consultation with those who had contributed to the professional literature: "others within the field." Other commentators were puzzling over the unexplained discrepancy between the prohibition on post–termination sexual relationships and the permissibility of other potentially harmful, exploitative relationships commencing only after termination. The emergent view seemed to be that both represented breaches of professional ethics.

Among the most common worries was that the psychologist, having come to understand a client's weaknesses during treatment, was in a position to take advantage of these vulnerabilities even years after treatment had ceased. The trustworthiness of practitioners would erode if psychologists were free to exploit their former patients as soon as therapy ended. Moreover, the former client might require a return to therapy, which could be blocked if a secondary relationship were in place. Concluding a comprehensive review of the literature on posttherapy non–sexual multiple relationships, psychologist Randolph B. Pipes proposed that the obligations of psychologists "should clearly prohibit us from unrestrained exploitation of recently terminated clients and that the profession should seriously consider limiting significant posttherapy nonsexual relationships between therapists and their former clients for at least some minimal period of time." (Pipes 1997)

My purpose in deposition was to show that a deeper and more comprehensive appreciation of professional ethics in psychology could reach well beyond the black letters of the code, that there were reasons for acknowledging a prohibition on exploitative and harmful nonsexual relationships commencing after treatment had terminated, that commentators on the profession had already appreciated the reasons for such an acknowledgement, and that the reasons for endorsing and conforming to such a norm were, in fact, good reasons. A responsible psychologist would never enter into such a relationship.

The parties to the lawsuit settled a few days after my deposition. The lawyers who were my clients were pleased with the outcome.

13.8 Reflections

The work of the expert ethics witness, as I have tried to understand and practice it, can make a contribution to the health of professions and their communities. For it is in courts of law where the reasonings of professionals and professions come under the most intense public criticism and review. Legal proceedings, like the one just described, are accountability rituals in which a profession's thinking about its social role can be publicly challenged. In a very real sense, the ultimate client of the expert ethics witness is the community as a whole. For a central collective task of a liberal profession is to identify and achieve a stable and just accommodation between those involved in the delivery of professional services and the members of the communities they serve. The task of reconciliation is rendered more difficult by moral pluralism and the tendency to act and judge out of personal interest rather than out of professional and civic values. Sound professional norms, and the practices that support them, can, if broadly honored by practitioners, enable the profession to discharge the social responsibilities assumed in the political process of professionalization, and enable the profession to secure and further its distinctive values exactly as citizens expect it to.

As I understand the task of the consulting ethicist, the expertise associated with this emerging role consists in (1) a deep and broad appreciation of the literatures on ethics in the professions, (2) the ability, over time, to resolve anomalies, contradictions and tensions within those literatures, and (3), the ability, consistently, to lay out sound reasonings in ways that the lay citizen can understand and appreciate. Today, ethics consultation is at best, only an embryonic profession. But it would be, I believe, a benefit to society if there were experts who could reliably serve in that capacity.

I do not ask anyone to take my word for anything. I am serving as what has been called a "knowledge broker," trying to provide my clients with what they need to know. Christine Korsgaard asked What is it that can sometimes endow ethical judgments with their remarkable power to impose themselves upon our decisions? The authority that counts is the authority we each have to reason through to our own conclusions. The ethics consultant possesses an Authority of Expertise. The normative authority that an ethics consultant can wield resides in the power, soundness, clarity, and applicability of the reasoned justifications that such a professional might make available. The work can remind us of the challenges and satisfactions to be found in mending a frayed social fabric, in healing a part of the world in a small but important way.

13.9 Postscript: The Samoan

As my medical ethics class wended its way through the syllabus, the case of the Samoan lingered with us. We began to ask "What should patients reasonably want a good doctor to care about?" making considerable progress towards reconciling the differences in personal morality that had compromised productive conversation at the beginning. At one point we explored the possibility of trimming informed consent to exclude Samoans, thereby respecting their culture: All adults should have the power to refuse medical treatments, except Samoans, who would have to follow parental directives. Remarkably, that approach satisfied no one. One of the Samoans, a young woman who was aiming for a career in medicine, was especially troubled. After an extended conversation in my office, she said "I guess it is hard to be a good doctor and a good Samoan." Months later I learned that she had given up her plan to become a physician and had returned to Samoa.

Suggested Readings

The selections listed below elaborate further on the themes developed in this article.

Kenneth, Kipnis. 1986. *Legal Ethics*. Prentice-Hall.

This a philosophical overview of the ethical dimension of legal practice.

Kenneth, Kipnis. 2012. Social Justice and Correctional Health Services. In *Medicine and Social Justice*, ed. Rosamond Rhodes, Peggy Battin and Anita Silvers. Oxford University Press.

Health care professionals working in prisons have obligations as doctors and nurses but also have separate duties as employees of the prison system. Who has responsibility for prison health care? How are conflicts between prison warden directives and professional obligations to be understood and managed?

Kenneth, Kipnis. 2008. Medical Confidentiality. In *Blackwell Guide to Medical Ethics*, ed. L. Francis, R. Rhodes and A. Silver. Blackwell Publishing.

An argument for a very strong obligation for medical confidentiality.

Kenneth, Kipnis. 2009. The Certified Clinical Ethics Consultant. In *HEC Forum*, 249–61.

Focusing on the embryonic field of clinical ethics consultation, an overview of the process of transforming an occupation into a profession.

References

American Psychological Association. 2010. *Ethical Principles of Psychologists and Code of Conduct*. http://www.apa.org/ethics/code/. Accessed 19 June 2016.

Cohen, G.A. 1996. Reason, Humanity, and the Moral Law. In *The Sources of Normativity*, ed. Christine M. Korsgaard, 167–188. Cambridge: Cambridge University Press.

Korsgaard, Christine M. 1996. *The Sources of Normativity*. Cambridge: Cambridge University Press.

O'Flaherty, Liam. 1961. *The Informer*. New York: New American Library.

Pipes, Randolph B. 1997. Nonsexual Relationships Between Psychotherapists and Their Former Clients: Obligations of Psychologists. *Ethics and Behavior* 7 (1): 27–41.

The manufacturer's authorised representative in the EU is Springer Nature Customer Service Centre GmbH, Europaplatz 3, 69115 Heidelberg, Germany. If you have any concerns regarding our products, please contact ProductSafety@springernature.com

Printed and bound by CPI Group (UK) Ltd, Croydon, CR0 4YY
22/09/2026
02221058-0010